Man-Machine Simulation Models

WILEY SERIES IN HUMAN FACTORS

Edited by David Meister

Human Factors in Quality Assurance

Douglas H. Harris and Frederick B. Chaney

Man-Machine Simulation Models

Arthur I. Siegel and J. Jay Wolf

Man-Machine Simulation Models

Psychosocial and Performance Interaction

Arthur I. Siegel and J. Jay Wolf

Applied Psychological Services, Inc.
Science Center, Wayne, Pennsylvania

Wiley-Interscience   A Division of John Wiley & Sons, Inc.
New York • London • Sidney • Toronto

To:

Estelle, Janet, Mark

and

Adele, Aron, Judith

PREFACE

The contemporary field of systems engineering has exhibited an increasing need for comparative evaluation of alternatives early in the research and evaluation phase. As the complexity and sophistication of systems increase, the rise in cost and time needed for their implementation makes more imperative the earliest possible prediction of their efficiency, thus enabling an early evaluation of modifications or alternative proposed systems.

A considerable body of applied techniques, useful in attacking these general alternative selection problems, has been developed in recent years by systems engineers. The large majority of the applications of these techniques, to date, have dealt only with equipment configuration and performance—with the underlying assumption that personnel were available or could readily be trained to control, operate, and maintain the equipment. Stated alternatively, the emphasis has been on selection of an optimum equipment system since man's capabilities seldom limited overall performance. This situation has begun to change in certain advanced equipment systems. "The missing or weak link in the planning of many expensive systems," admits de Sola Pool (1964) "is the prediction of human behavior." Fogel (1963) summarized the changing situation as follows:

"In the more recent past, however, increasing equipment complexity and environmental requirements have made it necessary for the designer to reach for the handbook to determine relevant data so that the human operator will "fit" the designed machine. We are presently faced with problems which require considerable understanding of the man and the machine so that both may be "married" in such a way that each one's attributes compensate for the other's deficiencies."

"Man-Machine Simulation Models" abstracts and presents the results of 10 years of research and digital experimentation conducted by Applied

Psychological Services, Inc. The work was directed toward the development of quantitative techniques for alternative system evaluation when personnel performance and interpersonnel relationships are conceded to be of importance to total system effectiveness—in fact, possibly the limiting aspect of system capabilities. It is a goal of the models described to predict system efficiency levels under various conditions that affect the performance of the man-machine system involved.

The initial impetus to the program was provided by the Office of Naval Research, Department of the Navy, in January 1958. This initial work yielded the two man-machine model described in the early chapters. Later, also under Office of Naval Research sponsorship, the techniques developed in the earlier work were amplified and applied to the development of a more sophisticated group-interactive simulation model which is described in later chapters.

ARTHUR I. SIEGEL

J. JAY WOLF

Wayne, Pennsylvania
May 1969

ACKNOWLEDGMENTS

We are deeply indebted to a number of associates for their advice, help and support over the years. Dr. Max Lund, Dr. Paul Cheatham, Dr. Richard Trumbull, Dr. Abraham Levine, Dr. James Miller, Dr. Marshall Farr, Mr. Gerald Malecki, and Mr. Luigi Petrullo, all of whom are currently or were previously associated with the Office of Naval Research, deserve the applause for any contributions that the techniques described may merit. We are also indebted to Mr. Allen Sjohelm at the Bureau of Naval Personnel and to Mr. James Jenkins of the Naval Ship Systems Command for the more recent opportunity to validate and expand certain aspects of the digital simulations. A large number of operational Navy establishments assisted in the collection of the criterion data reported throughout.

At Applied Psychological Services a host of persons provided invaluable assistance in a variety of ways. These persons included William Helsel, Gordon Bright, Richard O'Connor, Roland Desilets, Mary Knight, and James Andreas, who programmed the various models and who also contributed by insuring consistency between the various computational analysis and the equipment selected. Dr. Robert Smith, Mr. William Miehle, Mr. Ralph T. Sorenson, Mr. Richard S. Lanterman, and Mr. David Barcik contributed significantly to laboratory experimentation conducted for model validation, to the development of some of the psychosocial interactive digital simulation model concepts, and to field data collection and analysis. Mrs. Estelle Siegel served as program administrative officer throughout. She has somehow survived our spelling and performed the typing of the various project reports and the various drafts of this manuscript.

We acknowledge the substantial contributions of these persons and our indebtedness to them.

A. I. S.

J. J. W.

CONTENTS

xii   Contents

III.   RESULTS OF APPLICATION OF UNITARY OR
       DUAL OPERATOR SIMULATION MODEL   36

Carrier Landing   36
Missile Launching   40
Discussion of Landing and Launching   41
Inflight Refueling   42
Air Intercept   46
Conclusions   50

IV.   GROUP SIMULATION—QUALITATIVE AND
      CONCEPTUAL CONSIDERATIONS   52

Scope   52
Overview of Model   52
Major Variables in Small-Group Theory   54
Summary of Small-Group Variables   55
Norms and Goals—Orientation   55
Orientation of the Individual   57
Orientational Mix and Interdependence   58
Crew Orientation   58
Action Unit Orientation   59
Cohesiveness   59
Mission Orientation   60
Pressure and Reinforcement   60
Communication Network   62
Environmental Stress   62
Leadership, Roles, and Status   63
Group and Crew Size   63
Task Performance and Personality   63
Social-Cultural Characteristics   63

V.   THE GROUP SIMULATION MODEL—
     QUANTITATIVE CONSIDERATIONS   64

Brief Overview   64
Symbols and Limits   70
Detailed Model Logic   70
Mission Data   71
Equipment Data   72
Personnel Data   73
Parameters   74

I

MODELING AND SIMULATION

It is now becoming commonplace in scientific and business communities to predict gross outcomes of complex operations and event sequences in man-machine systems with the aid of the digital computer. The important features, characteristics, and outcomes of many systems, both real and planned, have been digitally reproduced by a process called "digital simulation." This method is proving to be a valuable aid to managers and analysts in such diverse disciplines as transportation, economics, international relations, population study, military operations, and logistics. The purpose of the technique is to provide quantitative estimates of the performance, efficiency, effectiveness, or "value" of systems or approaches so considered.

An example of the use of this technique is the simulation of a production plant to which a series of tasks is assigned. Here the computer is given data on the plant's resources—that is, the number of machines of each type, the number of men of each trade who can operate the machines, and the types of tasks to which they can be assigned. The computer is also given the workload that is to be processed through the plant during the time period to be simulated. This workload includes the work backlog at the beginning of the simulation and the tasks that are generated by a variety of means during the simulated period. Finally, the system operation rules are represented by a computer program of coded instructions. The program determines and controls such features as work-order priorities, overtime assignments, and methods for calculating how much, how fast, and how well the men and machines work when assigned to tasks in various combinations. Using all these data the computer processes the tasks sequentially in accordance with system operation rules by assigning them "on paper" (that is, in memory) to individual machines as if the tasks were actually routed from one machine to another. As part of the processing the computer maintains

1

records of interest such as how often a task must be delayed because of a need to wait for each machine, how the workload is distributed among the men, and the size of the backlog as a function of time and trade. These are summarized into totals, averages, and distributions which provide numerical measures of labor and machine utilization efficiency for determining how well the system performs its functions.

It is clear from this brief example that a rather thorough analysis of the total man-machine system and its assigned job must be made in preparation for simulation. Stated in terms applicable to digital simulation in general the total job is reduced to a list of events that may be initiated, performed, and terminated with stated frequencies and in specified sequences by preselected events or conditions. The resources of men and equipments that are available to accomplish the job are itemized. Their capabilities, limitations, and behavior are described numerically. These data together with selected parameter values stating general limiting conditions (for example, total time allotted and working-hour limits) are placed on some computer input media such as punched cards or magnetic tape. A computer program is developed that dictates the logic as well as the storage, arithmetic, and recording operation sequence according to the simulation rules. Only after all of these tasks have been completed can the computer accomplish the simulation by operating on the data in accordance with the programmed logic and produce tabular, textual, and/or graphic results for study and analysis.

Once prepared for automatic processing, a mission simulation can be repeated a number of times with variations in the load to be processed, the resources available, the decision logic to be implemented, the manning complement, speeds of equipment, level of equipment automation, and the changes of work-processing rules. It should be noted that the simulation technique does not purport to find the "best" solution to any problem; rather, it demonstrates the consequences of a particular set of input conditions and decision rules applied to a process. The simulation of a variety of conditions may facilitate predevelopment evaluation of complex systems or the effects of changes in existing equipments. Consequently digital simulation can be used to select among alternative system proposals. It is also possible that such a tradeoff analysis would facilitate discovery of potential difficulties such as equipment or personnel overloading which analytic treatment alone would not disclose. The digital approach to simulation and predevelopment system test can produce a host of system planning information not provided by other techniques.

MODELS

Despite the variety of current uses to which the word is put, it may accurately be said that this general technique, which has been described briefly, may be termed computer "modeling." Today the much-bandied term "model" has been so broadly defined as to incorporate virtually any form of abstraction used to represent concrete phenomena. Thus English and English (1958) under definition 4 state that a model is

". . . a description of a set of data in terms of a symbol or symbols, and the manipulation of the symbols according to the rules of the system. The resulting transformations are translated back into the language of the data, and the relationships discovered by the manipulations are compared with the empirical facts."

From this definition a theory and a model would seem to be interchangeable to the extent that both are symbolic. Chapanis (1961), and the present authors, would prefer a more stringent definition such that a model, serving as it does as an analogy, would be a representation or likeness ". . . of certain aspects of complex events, structures, or systems made by using symbols or objects that in some way resemble the thing being modeled." The "theory of relativity," then, would not be a model; yet Freud's "hydraulic model" of psychic forces depicted as impinging on consciousness or Descartes' mechanistic analogy of blood flow would seem to answer this criterion. Chapanis' companion distinction between "replica models" (for example, a wind tunnel), which resemble in a physical way the thing being modeled, and "symbolic models," which abstractedly represent the modeled event, aids in classifying models.

Another classification of models that helps to place computer simulation in its proper perspective is the view presented by Sayre and Crosson (1963). They consider three distinct ways in which an "object" might be modeled: (a) replication, (b) formalization, and (c) simulation. The first class includes physical reproductions, facsimiles, test models, duplications, and dummies. In the second category of formal models both the components of the system modeled and the interconnections among them are represented by symbols that can be manipulated according to the provisions of a formal discipline such as logic or mathematics. The result in this case is an analytic solution to a set of general equations. Models which fall into the third category are those whose equations or rules do not admit analytic solution in general form but that nevertheless produce numerical values given sufficient computing capacity and time. This latter class has been made practical in the recent past by the widespread availability and decreasing cost of high-speed electronic calculating equipment. It is to this category that the attention of a wide variety of recent model development has been focused and to which the present work is directed. For a definition of computer simulation model, per se, Martin (1968) proposes, "A logical-mathematical representation of a concept, system, or operation programmed for solution on a high-speed electronic computer."

The primary goal of all three types of models is, of course, the same— to produce the essence of the object modeled faster, at lesser cost, or with greater convenience than could be achieved with the object or system itself. The selection of the simulation model, in particular, to represent systems involving human behavior is a natural one. It is acknowledged that the assignment of a specific type of model to human-performance description is difficult, especially when the definition must be mathematical in nature. Human behavior with its associated idiosyncrasies is complex. This complexity along with the interactions among behavioral elements makes for large computational requirements. For this reason computer modeling seems quite useful for the simulation of human performance.

ATTRIBUTES OF SIMULATION MODELS

There are two primary qualities that it is hoped any predictive model, including simulation models, would possess. First, it is desired that the model will be sufficiently general to enable simulation of a reasonably wide class of systems and missions. With this property a relatively few different models would be required, lowering the developmental, programming, documentation, and test costs involved in modeling. However, in the present state of model development it must be stated, in candor, that relatively little attention has been paid to this desiratum. Generally one finds a variety of models developed, each one to simulate some specific situation: for example, the population projection of a country or the medical conditions of postnuclear attack. Yet within any individual model early attempts are being made to generalize at least within a given subject matter.

In an attempt to achieve this first quality of generality the development of a simulation model has been an evolutionary process. Sometimes it is only after a model has been used that additional desirable features are recognized. Improvements in the statistical processes involved, representation of variables, presentation format, numerical techniques, and operational utilization have often become evident through use and study. It also frequently happens that new features must be added to incorporate an aspect that is important for selected tasks. Thus, with use and age, the simulation model grows in capacity, flexibility, and generality. This is accomplished, however, only at the expense of complexity and often of longer computer running time.

The second desired quality is validity; that is, that the model predict actual performance of the simulated man-machine system within an acceptable margin of error. The extent to which this objective is achievable is overwhelmingly dependent on both the validity of data presented to the computer and on the accuracy of the assumptions that are employed in developing the model's logic. Cremeans (1968) in discussing advantages of digital simulation has stated the case plainly by admitting that "simulation in the face of poor data and weak theoretical foundation is only an advantage if we recognize the weakness of the outputs as well as the inputs."

For the time being those engaged in model development have concentrated on achieving a model as an admittedly rough tool to help solve a problem or at least to give insights to a solution. Currently the model-using community is groping to determine the limits of applicability of the new-found technique. The immediate objective seems to be to find something that works reasonably well and only then to extend it in generality and to determine and improve its accuracy.

It may be said that the predictive validity of many of these models has been shown to be generally acceptable. Certainly the appetites of those who have been involved have been whetted. They are anxious to state their limiting assumptions and caution against misuse of results. But they are also enthusiastic about learning how best to use this new tool. Whatever else may be said about accuracy, however, it is virtually unanimous that the systematization which the technique requires yields a wealth of information about the tasks to which they are applied.

Subsequent chapters present more explicit information indicating the level of success that may be expected with current digital simulation models.

SCOPE

This volume presents the bases, structural logic, and results of applying two distinct computer-based simulation models. The first of these attempts to model the activity of a one- or a two-operator man-machine system. This first model has been employed for such diverse simulations as landing an aircraft, firing an air-to-air missile, searching out, detecting, and classifying a hostile submarine, and reentering the atmosphere in a spacecraft. The second of these models is directed at larger systems and possesses the capacity for simulating the actions of dozens of men who work at several independent stations. Both of the models are used to represent systems composed of equipments that are monitored, controlled, or used by operators. In addition to the general limitations of such models, as discussed above, both of the current models assume that the sequence of operator actions is known and that within limitations the operators will act in a "lawful" manner. The construct validity of these models is limited by the theoretical substrate (status of human behavioral knowledge) on which they are based as well as by the various representations of this theory within the models. We note, however, that models are probably best judged on utilitarian and predictive validity bases rather than on the bases of their assumptions alone.

The major distinctive feature of these models, when compared with other digital simulation models, is their emphasis on psychologically oriented or psychosocial variables. In addition to the necessary but general aspects of the system and its resources that are incorporated into most simulation models several psychological aspects have also been incorporated into the current models. Along with the data on equipments, missions, work stations, and the simulated rules of conduct there are also included calculations of such variables as time-induced stress, proficiency, morale of the individual operators, as well as team decisions, cohesiveness, and psychosocial efficiency and orientation of a group. In addition to results on reliability and availability of equipments these models generate data on personnel performance. The total system effectiveness measures reflect both man and machine performances.

Both models are designed in agreement with the thesis of Bekey and Gerlough (1965) that "one of the most important aspects of digital simulation is the ability to handle stochastic events, that is, events which occur randomly. Techniques for such problems are often known as Monte Carlo methods."

In defining the operating model to serve as representation of certain relevant aspects of the psychological processes for computer manipulation it was agreed not to attempt to duplicate every minute aspect of a real life situation. Rather the simulations attempt to predict practical criteria by operating on a relatively small number of factors. These factors are believed to capture the essence of those variables that current leaders in the field of psychology agree upon as affecting the performance of the human. These factors, believed salient by others, are set into a form which allows digital computer simulation and are manipulated to predict the criterion situation.

The extent of the detail to be simulated always plagues the model builder. Too detailed a simulation must be expected to cost more and may only complicate the evaluation procedure. Too shallow a simulation may provide insufficient or misleading information. The extent of interaction of the variables has an important effect on computer storage requirements and consequently on the economics of computer time. In the case of the two models presented, a balance has resulted from consideration of alternative variables to be simulated during the early model-formulating stage.

In subsequent chapters the selection of variables included in the simulation model is presented and discussed for each model. These are woven into a framework of the logic of model operation and shown in the form of equations and logical flowcharts. The computational and programing aspects, a description of required input data, and the recorded output are given. Results are presented of tests and utilization of each model.

The model descriptions presented are designed to open a channel of communication between the psychologist and the engineer and scientist whose need for and interest in predictive estimates involving human performance are no longer questioned.

It is noted that in the present relatively early stage of simulation model development this work represents neither an ideal reflection of human performance nor a universally applicable or optimum computational implementation. It represents, rather, illustrative examples of the rationale and current status of digital simulation models which are the result of the application of several recently allied disciplines—individual and group psychological theory, high-speed digital computer technology, mathematics, and advanced man-machine systems planning. For a brief discussion of a variety of other diversified examples of the use of these techniques in digital simulation the reader is referred to Martin (1968).

WHEN SIMULATION IS APPROPRIATE

"Modern technology" emphasizes Apostel (1961), "utilizes a variety of models in the service of many different needs. The first requirement that a study of model building in science should satisfy is not to neglect this undeniable diversity, and to realize that the same instrument cannot perform all those functions." With this understanding the question arises: under what conditions can digital simulation models be expected to provide advantages over other approaches?

As a complement to specially developed simulation equipments or trainers, digital simulation provides a relatively inexpensive and timely method for yielding broad insights into the nature of operator performance in a man-machine system. If the real man-machine system is large, complex, or costly, the use of a computer for a predevelopment experiment may be inexpensive compared to the trial and error approach with the system itself; it often yields data for a larger number of cases than is practical by other approaches and at a lesser elapsed time per situation simulated. The same reasoning applies to the situation in which the system is in being but is so heavily occupied that experimentation with changes in equipment, personnel policy, or resources assignment rules may be impractical, expensive, dangerous, or unlawful. In these cases, too, available simulation techniques are appealing.

Considering the aspect of psychological impact on man-machine systems to which this volume is primarily addressed, it is by no means obvious that even carefully screened work teams are capable of performing adequately in the environment of some modern systems as conceived at this time. Prolonged confinement in relatively limited quarters (such as space vehicles, submarines, and airborne command posts), a hostile environment without, and the requirement to perform routine tasks, sometimes in communications isolation, combine to generate several military situations potentially limited by the operators rather than the equipment. Here again simulation is an attractive approach. Even when equipment is completed and available, the effort and expense of testing such systems operationally in its environment may be substantial and consideration of simulation in this instance to reduce such effort is also appropriate.

In summary, the principal value of digital simulation in its present state of development is that with it we can attack problems in advance of equipment production and study system loads and conditions of our choosing—conditions which may have otherwise yielded a less than optimal accomplished system.

II

LOGIC OF UNITARY
OR DUAL-OPERATOR
SIMULATION MODEL

SCOPE

This chapter presents and describes the logic and implementation of a digital computer model that simulates the actions of one or two operators who may work singly or together in the operation of an equipment. During the course of the description of the model the rationale for the selection of the particular model features is given from the psychological and human engineering points of view. The purpose of this model is to simulate numerically a man-machine mission which consists of a series of operator tasks, or equipment tasks, or both.

BACKGROUND

Pioneering work leading to the development of the unioperator version of this model was initiated by Applied Psychological Services early in 1958. The resulting computer program was prepared for the IBM 650 computer utilizing the SOAP (Symbolic Operating Assembly Program) technique. The model was applied to and tested through two naval tasks: landing an aircraft on the deck of a moving aircraft carrier and launching an air-to-air missile. Following corrections, improvements, and the addition of desirable features to the model and program, the general technique was expanded to the two-operator situation and reprogrammed for the IBM 705 computer using the Autocoder language. The model was tested and verified using the dual-operator tasks of inflight refueling of an aircraft and air-to-air intercept maneuvers. In addition two specially developed two-man synthetic tasks were developed. These tasks, for which criterion data are more readily attainable, were performed in the laboratory and were accomplished as well by a series of corresponding computer simulation runs for model verification. An exploratory application of an astronaut return-to-earth descent mission was also made. Most recently, again with improved features, the model was converted to the Fortran language for operation on the faster IBM 7094 system. Applications using this model have been accomplished for simulating sonar operation in submarines, a two-man space experimental laboratory mission, sonar maintenance, aircraft cabin evaluation, and segments of a lunar exploration mission.

It is this model in its latest form that is described in the present chapter.

INTRODUCTION TO MODEL

Within the advent of complex systems in which a human is expected to operate or control his machine it has become increasingly more likely to find during the test phase or in operational usage after the system is accomplished that the operator is overburdened or underburdened during the task. An after-the-fact remedy, if feasible at all, can be time consuming and expensive. This psychologically oriented model has been developed and refined in order to simulate the characteristics of the human operator in a man-machine system. Its goal is to allow a determination of where a man-machine system may overload or underload the operators while the system is in the early design stage, thereby avoiding the man-machine mismatch situation.

An assumption of major importance in the derivation of logic and concepts for the model is that this "operator loading" is the basic element in effective man-machine system performance. The fact that an operator may be overburdened or underburdened by a variety of causes, notwithstanding, this model concentrates most of these effected conditions into a variable called "stress." As used in the model an operator's increase in stress may be caused by several factors: (1) his own speed capability, that is, a stress buildup that results from falling behind in time on his assigned task sequence; (2) a realization that his partner, for any one of a number of reasons, is not performing adequately; (3) an inability to complete successfully the task aspects on the first attempt with the possible accompanying need to repeat these aspects; or (4) the need to wait for equipment reactions. During the running of computer simulations all of these conditions and a variety of other factors influence the stress variable. As stated by Appley and Trumbull (1968), given model inputs that include stressor dimensions and indices of a hypothetical man's adjustment potential and using differentially weighted response patterns as adjustment criteria, the model makes the stress variable the key to operator performance in terms of both speed and quality of performance.

PURPOSE OF MODEL

It is the purpose of this model to give the equipment designers quantitative answers to questions such as the following while the equipment is in the early design stage:

1. For a given operator procedure and a given machine design, can an average operator be expected to complete successfully all actions required in the performance of the task within a given time limit?
2. How does success probability change for slower or faster operators and longer or shorter periods of allotted time?
3. How great a stress is placed on each operator during his performance and in which portions of the task is he overloaded or underloaded?
4. What is the distribution of operator failures as a function of operator stress tolerance and operator speeds?

OVERVIEW OF MODEL

Prior to the use of the model a task analysis is performed for the man-machine mission under consideration. For each operator action to be performed, called a "subtask," certain specific required source data are compiled for each operator. These data, together with the parameter values selected, are prepared in punched card form and introduced into the digital computer for which a computer program has previously been prepared. As directed by its program the computer starts at time zero and simulates the performance of each subtask by the operator in the proper sequence according to the rules of the model. Operating on the source data the computer simulates the performance of each operator by calculating values for and keeping track of items such as the operator's subtask execution time, failures, stress, and idle time for each subtask and for each operator. Since stress is based on time pressure, the model is largely, but not exclusively, time oriented. Subtask execution time is stochastically determined from specified normal distributions using a Monte Carlo techinique and is dependent upon stress levels as well as the operator speed parameter. Subtask success or failure is also stochastically determined on the basis of performance, stress, and probabilities supplied as input. The following important items are also calculated or considered in the model:

- Subtask precedence (variable sequencing of subtasks)
- Operator interaction (waiting for a partner)
- Joint subtasks (performed by both operators simultaneously)
- Equipment delays
- Time precedence (idling until a given time occurs)
- Operator decisions
- Essential versus nonessential subtasks
- Operator cohesiveness
- Time spent waiting for a prespecified event

Several of these concepts are also implemented in the model by the use of Monte Carlo or game-of-chance techniques in which random samples are chosen from probability distributions to determine step-by-step function values and to select alternative courses of action. This nonanalytical approach results in approximate solutions whose accuracy is dependent on the number of iterations or samples taken. The technique is based on the statistical premise that the expected result of a given performance however complicated, can in principle be estimated by averaging the results of a larger number of individual attempts, each one of which is determined more efficiently by another simulated performance known to have the same expected result. Each operator action taken separately can be said to have an inherent predictability under ideal conditions.

The simulation itself, consisting of the calculation of values for variables, determination of subtask success, and the proper bookkeeping, continues serially for each subtask. A simulation is completed when the simulated operators either run out of allotted time or successfully complete the task within the available time. Following the computer's "performance" of a task, a set of output data is recorded by the computer to indicate the areas of operator overload, failures, idle time, subtasks ignored, and the like, for the given set of selected parameters. The total task simulation performed is repetitive since, in order to simulate intra- and interindividual differences in performance the simulation of any individual task is based, in part, on a random process. Just as subtasks are simulated sequentially to comprise a total task, each task is repeated many times with a fixed set of parameters to obtain averages of the data generated by randomization techniques.

Further repetitions of the mission with different parameter values yield additional printed listings consisting of tables, frequency distributions, and graphs. Analysis of the results yields insights to the mission,

its composition, and operator performance. Often if alternative equipment designs or task sequences are indicated, they are similarly prepared, simulated, and analyzed in order to determine the extent of
improvement they may yield.

The program offers a variety of options to the user relative to examining different equipment and personnel characteristics. There is no theoretical limit to the type or detail of the task that may be simulated.
A variety of information options are offered the user so that in addition
to the final results he can obtain intermediate results that permit study
of individual subtask performance, time periods, and function values
which can be utilized for better understanding of the equipment system
under analysis.

INPUT DATA

To use the model the following seventeen items of task analytic input
data are required for each subtask and for each operator. Insofar as
possible a subtask is selected to be a "natural" behavioral unit; usually
this is a single operator action, although it can be two or more actions
that normally occur together. In most tasks to which the model has
been applied subtasks have been selected to represent actions that can
be accomplished in from several seconds to a few minutes. Thus current
applications are to missions of duration of less than one hour, although
this is not a model limitation. These data may be derived from such
procedures as task analysis, formal experiments, informal measurements,
literature search, or personal interviews. The required input data to
be punched on subtask cards for each task and operator are:

1. *Operator number:* $j = 1$ or 2, identifies the operator who is assigned
to the subtask.

2. *Subtask number:* i, an integer that identifies the assigned subtask.

3. *Type of subtask:* a code indicating one of four special subtask types.
Any type can appear without restriction wherever desired in the task
sequence. A joint subtask (type $= J$) is one performed simultaneously
by both operators; for example, a communication task is simulated simultaneously with one operator talking and one listening. (A separate card
is provided for each.) An equipment subtask (type $= E$) is introduced
to account for a delay in the task because of factors other than human
performance (for example, to simulate an equipment warmup). No operator stress functions are calculated for this type of subtask. A decision
subtask (type $= D$) is incorporated into the sequence to cause branching, skipping, or looping in the task sequence to simulate a choice made

by an operator without the operator taking any action. A cyclic subtask (type $= C$) requires an operator to wait until the start of the next P-second periodic time interval before he can initiate the subtask (P is given as a parameter). A blank is indicated for a regular action subtask in which an operator performs some motor or mental effort.

4. *Indication of subtask essentiality:* an indicator specifying whether or not the successful performance of the subtask is essential to successful completion of the task. This datum allows the computer to identify and ignore nonessential subtasks during "highly urgent" conditions. (E = essential; N = nonessential).

5. *Subtask precedence:* d_{ij} (mnemonic delay): a number indicating a subtask that must be successfully completed by his partner before an operator can begin the current subtask. By proper selection of d_{ij} values it is possible to cause either operator to "wait" until his partner has completed a stipulated subtask successfully. Thus "waiting" for one's partner is simulated differently from time spent "idling" until a fixed time as in item 6 below.

6. *Time precedence:* I_{ij}: the point in time before which operator j is not permitted to begin subtask i.

7. *Next subtask, success:* $(i,j)_s$: the subtask to be performed next by operator j if he succeeds on subtask i or if he selects the first alternative course in a decision subtask.

8. *Next subtask, failure:* $(i,j)_f$: the subtask to be performed next by operator j if he fails at subtask i or if he chooses the second of two alternative courses in a decision subtask.

9. *Average subtask execution time:* $\bar{t}_{ij}$: the average time required by the jth operator to perform subtask i. This average value represents the case in which the operator is under no stress. Examples of values suggested as tentatively applicable for representative subtasks are shown in Table 2.1. The average execution time data presented in Table 2.1 were drawn from a variety of sources. Principal among the sources were human factors experiments which were devoted to measuring control activation times. In some cases several different sources reporting differing activation times for seemingly similar controls were located. In these cases mean response time and variance were calculated. The user of the model may find that certain responses for some of the actions with which he is concerned are not available in the literature or in various data stores. We have found that informal stop-watch trials provide sufficiently sensitive execution time data for employment in the model. We also note that the subtask execution times contained in Table 2.1 are representative of field conditions. Under highly controlled laboratory conditions the values of Table 2.1 should be reduced by 50 percent

because the distractions and variability of operations are not usually present under sterile laboratory conditions.

10. *Average standard deviation:* $\bar{\sigma}_{ij}$: taken around the mean $\bar{t}_{ij}$ for the average operator while not under stress. Examples of values for these data are also shown in Table 2.1.

Table 2.1   Examples of Average Execution Times and Standard Deviations for Representative Operator Actions *

Operator Action	Average Execution Time $\bar{t}_{ij}$ (sec)	Average Standard Deviation $\bar{\sigma}_{ij}$ (sec)
Set toggle switch	1.1	0.76
Set rotary control	8.6	3.00
Push button (or foot switch)	4.2	1.02
Lever (throttle) setting	3.0	0.48
Joystick setting	3.8	0.48
Read n instruments	$0.6n + 0.6$	$0.2n + 0.2$
Communication, n words	$0.66n + 0.6$	$0.34n + 0.4$
Ignore nonessential subtask when situation is "highly urgent"	0.6	0.0

* Each $\bar{t}_{ij}$ value includes 0.6 sec to allow for attention shift between subtasks and, similarly, each $\bar{\sigma}_{ij}$ includes 0.2 sec.

11. *Average subtask probability of success:* $\bar{p}_{ij}$: the probability that the average operator j while not under stress can perform subtask i successfully or that he will select one or another course of action in a decision subtask. The assignment of subtask success probabilities can be made on several bases. We have relied largely on logic, a knowledge of the characteristics of the subtasks under consideration, informal observations, and interviews with systems operators. A number of data stores are becoming available and these will eventually be useful for this application. For most subtasks probabilities of 0.97 and above have been found to be appropriate.

12. *Time remaining, essential:* $T_{ij}{}^{E}$: the time required to perform all remaining essential subtasks (including i) at average execution times, assuming no failures.

13. *Time remaining, nonessential:* $T_{ij}{}^{N}$: the time required to perform all remaining nonessential subtasks (including i) at average execution times, assuming no failures.

14. *Indication of two special subtask types:* the allowance for one operator to make a decision that will decide the sequence of future

subtasks for both operators. The first of these enables each operator to jump to an individually specified subtask if one of the operators ignores the special jump subtask type 1. If a subtask so identified is ignored because of stress levels, operator j will go to the special jump subtask as indicated in item 15 below for his next subtask and his partner will go to the special jump subtask as indicated in item 16 below. If the task is not ignored, $(i,j)_s$ and $(i,j)_f$ will apply as usual. Special jump subtask type 2 provides a team decision capability to the model. If a subtask so identified is a success (probability $= \bar{p}_{ij}$), then operator j will go to $(i,j)_s$ for his next subtask. If the subtask is failed, however, the current operator goes to the subtask indicated in item 15 and his partner goes to the subtask indicated in item 16. In these ways one operator can make a decision that will determine the future sequence of subtasks for both operators.

15, 16. *Next task numbers:* like 7 and 8 above for use on special subtasks.

17. *Goal aspiration:* $G_{i,j}$: the perfomance level at which operator j is satisfied with his performance on subtask i. Use of this datum, given on a zero-to-one scale, is optional.

PARAMETERS AND INITIAL CONDITIONS

The other data required by the computer in advance of the simulation are the parameters and the initial conditions. These permit the adjustment of critical variables and the consequent determination of the range of their effects. There are four principal pairs of parameter values specified for each run. These pairs, one for each operator, provide the flexibility to vary run conditions for the following: mission time limit, T_j; operator stress threshold, M_j; operator individuality factor, F_j; period for cyclic subtasks, P_j.

The parameter T_j specifies the total times allotted to each operator for performance of the task. For a two-man team the task is considered to have been successfully completed only if both operators complete all required subtasks within the time specified by their respective time limits.

The parameter F_j, which accounts for variance among individuals, is the individuality factor for each operator. It provides the ability to simulate an operator who usually performs faster or slower than the average operator for whom an F_j value of unity is assigned. The effects of faster, or more highly motivated operators $(F_j < 1)$, and slower operators $(F_j > 1)$ in the performance of the task are examined by performing several computer runs with different F_j values. The range of values for F_j from 0.7 to 1.3 has been found to be practically useful in simulations.

A primary assumption in this model is that the "certainty" in the operator's "mind" that there is insufficient time remaining to complete the essential subtasks when performing at normal speed and efficiency will cause a state of stress on the operator. The model defines stress (s_{ij}) as a state of mind of operator j just prior to his initiation of subtask i. Current psychological theory suggests that emotion or stress up to a certain point acts as an organizing agent. Accordingly initial stress buildup is recognized in the model as having an organizing effect on operator performance as long as the value of stress remains less than the stress threshold parameter M_j. When stress exceeds M_j, the effect is disorganizing. The stress function, presented analytically later, may briefly be described as the ratio of how much is left to do to the amount of time available in which to do it. Thus, the stress threshold may be considered as the operator's breaking point. For example, an M_j value of 2 indicates that the operator begins to become slower and less accurate at the point at which he has more than twice as much work to do (at average speed) as he has time remaining in which to do it. Prior to this point any added backlog of essential subtasks creates a mental inducement of stress that makes his actions faster and more accurate. This effect is shown in Figure 2.1. Values for M_j in the range of 1.9 to 2.8 have received primary attention and represent the range of M_j values over which the model has been found to possess predictive pre-

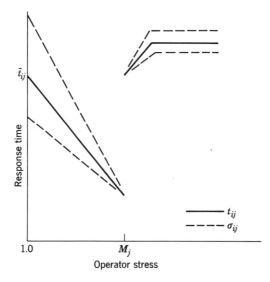

Figure 2.1   Operator response time as a function of time stress.

cision. Originally the M_j function was derived by the authors on the basis of their understanding of Cannon-Bard theory, which views mild stress as organizing and severe stress as disorganizing. Later (Siegel, Wolf, and Sorenson, 1962) an experiment was performed to calibrate the function. The alternate function, shown later in Figure 2.3, was derived as a result of this experiment. An M_j value of 2.3 has been found to be appropriate for the "average" person. Persons who are "more calm" or "less calm" than average have been simulated with values between 1.9 and 2.2 or between 2.4 and 2.8, respectively. The stress concept is implemented within the model by reducing response times and standard deviations up to a given point (the M_j value) and then increasing these values to simulate disorganization. Haythorne and Altman (1967) present a curve based on two separate experiments that supports the first half of the stress function. These authors go on to say,

". . . Thus, the combined findings suggest that, up to a point, increased levels of stress are associated with enhanced functioning. Beyond that, elevated stress may be associated with degraded performance output. In the present study, it appears that as the number of sources of stress was increased, . . . , a point was approached where the enhancing effect of stress on performance began to level off. Had our situation gone further, in terms of induced and experienced stress levels, performance might have begun to deteriorate."

The fourth parameter pair represents a time period P_j, that is applicable only to cyclic (type C) subtasks. A class of subtasks has been encountered at various times during the course of application of the model involving actions that cannot be initiated by the operator until an equipment-related periodic time interval elapses. Such tasks are found, for example, in work with radar or sonar systems in which the scan time —specifically the time of the beginning of a scan—is the critical event before which an operator cannot initiate his subtask. In these cases the equipment scan time is constant and the effect is a periodic one. When such a subtask occurs in the task sequence, the simulated operator must wait until the start of the next period before he can begin his subtask. The period P_j is given as a parameter for each operator.

Because the simulation of any individual task is based in part on a random process, it is necessary to repeat the simulation of a task many times in order to obtain sufficiently representative performance data for each set of conditions. Consequently, prior to simulation a selection is made for the value of the parameter N to indicate the number of times a given task is to be simulated; thus there are said to be N simulations, or N iterations, per computer run. According to the statistical law of large numbers the results tend to stabilize as the number of iterations is increased. Implementation of this selection in the model is in accord with the view expressed by Bekey and Gerlough (1965),

"Statistical variations between successive trials of the same experiment are also unavoidable in systems where human beings perform or operate as parts of the system. In such cases human beings must also be included as portions of simulated systems and reliable results can be obtained only if large numbers of simulated experiments are conducted in order to take into account the statistical variability of human performance."

Tests using the model have verified this tendency and indicate that a value of N from 100–200 is a satisfactory range for stable results.

Another initial condition of the model is R_0, the number from which the computer generates subsequent pseudo-random numbers R_1, R_2, R_3, . . . , needed during the course of the simulation. These numbers are termed pseudo-random because they are generated by a repeated nonrandom process but the result is indistinguishable from numbers that result from a truly random or stochastic process by reasonable statistical tests. Unless otherwise specified the last number generated in one run is used as the first value in the next run. Although individual task performance simulations are based on pseudo-random numbers, the fact that the initial number of the series is specified as a parameter enables any given computer run or series of runs to be repeated.

The following two examples illustrate the principal uses of pseudo-random numbers in the calculations:

1. To determine alternative courses of action: given the probabilities of the alternatives which sum to unity a pseudo-random number equiprobable in the range 0–1, will determine to which action it corresponds.

2. To determine the value for a variable known or assumed to have a given statistical distribution (e.g., rectangular, Poisson, normal): given the principal parameters of the distribution and one or more such pseudo-random numbers appropriate arithmetic algorithms are available for operating on them to select a sample value.

Other important initial conditions of the model are as follows:

1. A run number for run identification
2. A unit time code to specify the output in units of seconds, minutes, or hours
3. An indicator to specify whether or not the goal orientation variable is to be utilized

For a given run, then, N computationally repeatable task simulation iterations are made with a fixed value for each of the parameters M_j, T_j, F_j, and P_j, starting with R_0 as the first pseudo-random number.

THE SIMULATION SEQUENCE

The general computation sequence for the simulation model is shown in flowchart form in Figure 2.2. During the course of the explanation of model logic reference is made to the flowchart and the composite list of model symbols in Appendix A. Following the initial operation, namely, the reading of its coded instructions, the task analytic data are read and stored. A run begins when parameters and initial conditions are read and stored and a variety of bookkeeping functions are accomplished to set (or reset) the computer to perform the first operator's initial subtask. The determination of which operator to simulate at any given time in the sequence depends on $T_{ij}{}^U$, the total time used by operator j in "performing" all subtasks from the start of the simulation through his most recently completed subtask. The operator having the smaller value is selected and his next subtask is simulated. Subtasks are simulated in the sequence determined by the $(i,j)_s$ and $(i,j)_f$ input data and by a Monte Carlo determination of task success or failure described later.

WAITING AND IDLING

If the operator so selected must wait for his partner (as specified by d_{ij}), however, the sequence continues using data for the other operator. Next, a determination is made as to whether the operator must idle until an amount of time I_{ij} has elapsed from the beginning of the simulation. If idling is required, the idle time $(I_{ij} - T_{ij}{}^U)$ is recorded, total accumulated, $T_{ij}{}^U$ set equal to I_{ij}, and the control returned to determine which operator to simulate next. If no idling is required, a determination is made as to whether or not subtask i is a joint (type J) subtask. If it is, the operators are synchronized by setting the total time used by both to the value of the one who has used more time. This may result in a wait for either operator and is treated as the wait described above.

URGENCY AND STRESS

Except for decision and equipment subtasks, one of three states of "urgency" is determined for all subtasks for each operator. Urgency

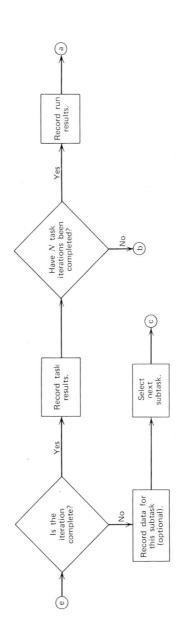

Figure 2.2 General flow chart of the one or two operator man-machine model.

21

is based on the remaining time available to an operator for completing the task, the operator's individuality factor, and the average time required to complete the task if no failures occur:

- The situation is *nonurgent* when sufficient time remains to complete all remaining subtasks.
- The *urgent* state occurs if the time available is insufficient to complete all remaining subtasks but is sufficient for completing all remaining essential subtasks.
- The situation is *highly urgent* if there is insufficient time available for completing even the remaining essential subtasks.

In the urgent or the highly urgent conditions the computer ignores nonessential subtasks.

Following the determination of the degree of "urgency," the stress condition is calculated. During nonurgent and urgent conditions, s_{ij} is defined to be equal to unity. When the situation is highly urgent, stress is defined as the ratio of the sum of the average execution times for the remaining essential subtasks to the total time remaining:

$$s_{ij} = \frac{\bar{T}_{ij}{}^{E}}{T_j - T_{ij}{}^{U}}$$

where T_j is the total time available and $T_{ij}{}^{U}$ is the time used up to, but not including, subtask i. The value of stress calculated is limited by the program to lie between one and five. The impact of stress on other variables is pronounced, particularly on execution time and subtask success as will be described.

Since each operator has an individual time limit on his performance and a task failure occurs only when the larger of these limits is exceeded, it is possible for the simulation to continue with one operator having exceeded his time limit. Should this be the case, the stress condition of this operator is set equal to his stress threshold value M_j, for the remainder of the simulation.

TEAM COHESIVENESS

The model also simulates each operator's confidence in, or cohesiveness with, his partner. Lack of team cohesiveness may reflect disagreements about goals or their importance or about methods or locus of authority. An operator can often determine how well his partner is performing. When one operator sees, hears, or otherwise "feels" that his partner is not performing satisfactorily, he will probably modify his own actions. On the other hand the "peace of mind" resulting from adequate partner performance permits an operator to perform normally;

that is, no change in his own stress occurs. The model accomplishes this by incrementing the stress value for an operator if his partner has a stress value greater than unity. The stress additive A_{ij} is calculated as follows:

$$A_{ij} = \begin{cases} 0 & \text{if } s_{ij'} = 1, & \text{no partner stress} \\ \dfrac{s_{ij'} - 1}{M_{j'} - 1} & \text{if } 1 < s_{ij'} \leq M_{j'} & \text{moderate partner stress} \\ 1 & \text{if } s_{ij'} > M_{j'} & \text{excessive partner stress} \end{cases}$$

where j' denotes the partner. A new stress value $A_{ij} + s_{ij} = S_{ij}$ is used in later calculations of subtask performance time.

An index of cohesiveness C_{ij} is also calculated for each operator on each subtask as a measure of the joint stress condition of the team. Cohesiveness is the product of the stress levels of the two operators normalized by their respective stress threshold values:

$$C_{ij} = \frac{s_{ij}s_{ij'} - 1}{M_j M_{j'} - 1}$$

As a result when neither operator is under stress, $C_{ij} = 0$, and when the stress on both operators is equal to their respective thresholds, $C_{ij} = 1$. Thus increasing C_{ij} values indicate greater team disharmony.

CYCLIC SUBTASKS

Before calculation of the subtask execution time for cyclic subtasks, the time to the beginning of the next P-sec period is determined and the cumulative time used for all preceding subtasks and waits is adjusted. First, the time used is divided by the period P_j. If the quotient is an integer, the operator is just ready to begin a P-sec period and the subtask can proceed immediately with the calculation of execution time. If the quotient is not an exact integer, however, the start time of the subtask is determined to be $P_j(1 + [\text{quotient}])$ where the [quotient] is the integral part of the division. The difference between this starting time and the previous cumulative time used $T_{ij}{}^U$ is counted as cyclic waiting time spent by the operator in waiting for the P-second period to begin.

SUBTASK EXECUTION TIME

Next, the time that the simulated operator takes to complete the subtask is computed. The purpose of the logic and formulas for subtask execution time is to provide a value selected from a truncated normal distribution in which the input values of $\bar{t}_{ij}$ and $\bar{\sigma}_{ij}$ are

- Used unchanged when augmented stress equals unity.
- Decreased in accordance with an empirically determined cubic function with increasing augmented stress until the threshold value is reached.
- Used unchanged when augmented stress equals the threshold value.
- Increased linearly with increasing stress beyond the threshold until when augmented stress equals $M_j + 1$, the contributions of $\bar{t}_{ij}$ and $\bar{\sigma}_{ij}$ remain constant at $2\bar{t}_{ij}$ and $3\bar{\sigma}_{ij}$ respectively.

These functions are shown graphically in Figure 2.3.

The average operator ($F_j = 1$) will require t_{ij} sec to perform subtask i when $S_{ij} = 1$. In this case his average standard deviation will be $\bar{\sigma}_{ij}$. Of course, no two operators would be expected to perform any subtask in exactly the same time on each repetition and no operator would be expected to perform the same subtask identically on two occasions except by chance. Therefore specific values for the actual subtask execution t_{ij} are selected by a Monte Carlo technique from a normal distribution limited from below by a fixed minimum, selected as 0.75 sec. Two pseudo-random numbers R_1 and R_2 uniformly distributed in the unit interval are sequentially generated by the computer from the initial value of R_0. From these numbers corresponding values of an independent, random, normally distributed variable K_{ij} are generated. Thus if K_{ij} is the number of standard deviations from the

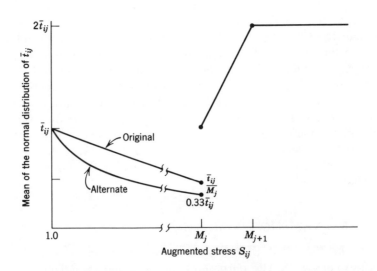

Figure 2.3   Average execution time as a function of stress.

mean corresponding to the pseudo-random numbers generated, then t_{ij} which implements the desired relationships is given in the original scheme by:

$$t_{ij} = \begin{cases} \dfrac{V_{ij}F_j}{S_{ij}} & \text{if } S_{ij} < M_j \\ [(2S_{ij} + 1 - 2M_j)V_{ij} - (S_{ij} - M_j)\bar{t}_{ij}]F_j & \text{if } M_j \leq S_{ij} \leq M_j + 1 \\ [3V_{ij} - \bar{t}_{ij}]F_j & \text{if } S_{ij} > M_j + 1 \end{cases}$$

where $V_{ij} = \bar{t}_{ij} + K_{ij}\bar{\sigma}_{ij}$ and K_{ij} is a random deviate—that is, a pseudo-random number selected from a normal distribution with mean of zero and unit standard deviation.

An alternate function for t_{ij} effective only in the region in which the augmented stress of the operator is below his threshold has been derived. The revised execution time is:

$$t_{ij} = F_j V_{ij} Z_{ij} \quad \text{if } S_{ij} < M_j$$

$$Z_{ij} = -1.829 \left(\frac{S_{ij} - 1}{M_j - 1}\right)^3 + 3.472 \left(\frac{S_{ij} - 1}{M_j - 1}\right)^2 - 2.350 \frac{S_{ij} - 1}{M_j - 1} + 1.0$$

An equipment (type E) subtask provides for the case in which a delay is introduced and time is used because of factors other than operator performance. The technique for determining the execution time in this case is similar to the calculation of t_{ij} as discussed but without the effects of operator stress involved (that is, Z_{ij} is set equal to unity).

SUBTASK SUCCESS AND FAILURE

The model assumes that the probability of operator success on a subtask increases linearly with stress from a value of $\bar{p}_{ij}$ (provided as subtask input) until it assumes a value of unity (certainty) at the stress threshold. At this point the probability assumes the average value $\bar{p}_{ij}$, after which it decreases linearly until, when stress has a value equal to $M_j + 1$, it levels off at a value that is decreased from $\bar{p}_{ij}$ by an amount equal to $1 - \bar{p}_{ij}$. To accomplish this the actual probability of successful performance of a given subtask p_{ij} is given as a function of $\bar{p}_{ij}$, s_{ij}, and M_j, as follows:

$$p_{ij} = \begin{cases} \bar{p}_{ij} + \dfrac{(1 + \bar{p}_{ij})(s_{ij} - 1)}{M_j - 1} & \text{if } s_{ij} < M_j \\ \bar{p}_{ij}(s_{ij} + 1 - M_j) + (M_j - s_{ij}) & \text{if } M_j \leq s_{ij} \leq M_j + 1 \\ 2\bar{p}_{ij} - 1 & \text{if } s_{ij} > M_j + 1 \end{cases}$$

In order to determine actual success or failure for any subtask the computer generates a pseudo-random number R_3 uniformly distributed over the unit interval. The operator is considered to have performed the task successfully if R_3 is less than p_{ij}; otherwise he is assumed to have failed. This implies that in the long run there will be a failure with probability p_{ij}.

To facilitate the calculation these expressions were rearranged to indicate success if:

$$\frac{(M_j - 1)R_3 - s_{ij} + 1}{M_j - s_{ij}} < \bar{p}_{ij} \qquad \text{when } s_{ij} < M_j$$

$$\frac{s_{ij} - M_j + R_3}{s_{ij} - M_j + 1} < \bar{p}_{ij} \qquad \text{when } M_j \leq s_{ij} \leq M_j + 1$$

$$\frac{R_3 + 1}{2} < \bar{p}_{ij} \qquad \text{when } s_{ij} > M_j + 1$$

In event of either success or failure input information indicates the task that is performed next.

GOAL ASPIRATION

The model provides for each operator's goal aspiration to affect his simulated performance. As the term is employed here goal aspiration refers to the goal setting behavior of an operator and the performance facilitation or debilitation that attends his goal getting. In essence goal aspiration is viewed as a level of aspiration phenomenon. The specific manifestations that are simulated are each operator's perception of the level of subtask performance success he expects to achieve, each operator's continual readjustment of this expectancy level on the basis of how well he is performing, and his stress.

Goal setting, aspiration to goal achievement, and the interaction of these factors with performance previously have been shown to be fundamental determinants of behavior. The goals an individual sets for himself determine how well he will perform and, in turn, his performance considerably influences the goals he sets. Thus for a given subtask it is not unusual to observe an individual setting a goal for himself that does not correspond perfectly with his performance capability (the goal may be higher or lower, depending on a number of factors). Moreover the individual will be more or less effective in his next performance of similar and subsequent subtasks, depending on whether the goal is achieved or not. As expressed by Lewin (1942), "A successful individual typically sets his next goal . . . above his last achievement. . . . The unsuccessful individual, on the other hand, . . . becomes intimidated and gives up reaching out toward higher goals. . . ."

The value of goal aspiration G_{ij} given as subtask input data is viewed as a numerical measure of the self-orientation of the operator; it is the level at which he is satisfied with his own performance. In implementing the feature a value is calculated for goal discrepancy on each subtask except the first ten. It is the difference between G_{ij} and the operator's current performance level, where performance is defined to be the ratio of the number of subtask successes to the total number of attempted subtasks. Any of five situations may result, as shown by Table 2.2. If the difference is small (from −0.02 to +0.02), no adjustment is made. Case I presents the circumstances of the performance level being lower than the goal level and the operator not incurring severe stress. Under such circumstances the real life expectation is that the operator will be more motivated to attain his goal. This motivation is reflected through "working better" and is simulated by increasing the probability that he will succeed in his subsequent effort. In case II when performance exceeds the goal level and the operator is not severely stressed, he would be expected to establish a new and higher goal for the quality of his future performance. For example, Deitsch (1954) wrote, ". . . most people of western culture, under the pervasive cultural pressures toward 'self-improvement' . . . tend to keep their level of aspiration higher than their previous performance." This is simulated by adding to the operator's G_{ij} value in accordance with a stochastic pro-

Table 2.2   Program Actions as a Function of Goal Discrepancy and Stress

Case	Goal Discrepancy (G_{ij} − Current Performance)	Current Stress	Program Action
0	−0.02 to +0.02	Any value	No action
I	Positive (>0.02)	Equal to or below threshold	Increase probability of success on next subtask
II	Negative (<−0.02)	Equal to or below threshold	Increase goal aspiration for next subtask
III	Positive (>0.02)	Above threshold	Reduce goal for next subtask to current performance level and reduce probability of success on next subtask
IV	Negative (<−0.02)	Above threshold	Reduce current stress to stress threshold

cess. The value for the increase is selected to be $0.05 \pm 0.03K$, where K is a different normal deviate value. A minimum increase is selected to be 0.01.

When severely stressed and performing below his goal level, as in case III, the normal individual has no choice but to lower his goal level. Because of the high stress he is incurring, he must accept his current performance as satisfactory. In essence this is a point of resignation to "about the best I can do." However, attending the point at which the individual ceases his upward striving is a deterioration of performance. This is represented in the simulation in this case by reducing the probability of success associated with the ensuing subtask.

Case IV involves the circumstance in which the operator is under severe stress at a time when his performance equals or exceeds his goal. In such a circumstance an operator might ask himself, "Why try so hard?" The current stress level is reset by program action from supraliminal back to his stress threshold. When on the next subtask his stress is reduced to the threshold and if at that time performance is still in excess of the goal, his circumstances now become subject to the rule for case II. This dynamic interaction of effect and cause should be noted for all cases so that the action of a given case may render the operator subject to the considerations of one of the other three cases on the next subtask. In this way the dynamic nature of cause and effect within individual goal setting, testing, adjusting, resetting, retesting, readjusting, and so forth, is represented in the simulation model.

Success probabilities for cases I and III for the next subtask are calculated as $\bar{p}_{ij}[1 \pm 0.08 \text{ (goal discrepancy)}]$. The subtractive process is performed when current stress is equal to or below the stress threshold.

DECISIONS AND SPECIAL SUBTASKS

The remaining computer operations are concerned with bookkeeping and updating of memory values, prior to recording of results. However, prior to a discussion of the form and content of the results, several special subtasks and applicable techniques will be discussed.

DECISION SUBTASKS

Operator decisions during the performance of a task represent an important feature of the model in agreement with the opinion of Pew (1965) that "More and more frequently the tasks assigned to men in systems involve information processing in situations in which speed of processing is not the most important dimension, but the quality of the decision is." The purpose of this type of subtask is to simulate the real world by providing for possible subtask execution in other than a straight, linear sequence. For example, an operator may find it desirable or external conditions may require him to skip one or more subtasks. Or having reached a critical point the operator may be faced with several alternative courses of action. The decision subtask is incorporated to enable such branching, skipping, and looping. It causes the computer to select the next subtask without consuming simulated operator time. No operator time is counted for decision subtasks since the time to shift attention between subtasks is included in the time for each subtask. Decision subtasks may be placed anywhere in the sequence. For these $\bar{t}_{ij}$, $\bar{\sigma}_{ij}$, and essentiality have no meaning as the calculation of execution time is by-passed. To determine the next subtask another pseudo-random number is compared against the $\bar{p}_{ij}$ of the decision subtask. Therefore the next subtask to be performed as a result of the decision will be subtask $(i,j)_s$ with probability $\bar{p}_{ij}$ or subtask $(i,j)_f$ with probability $(1 - \bar{p}_{ij})$.

MULTIPLE-ACTION SUBTASKS

In certain subtasks, such as in manually controlled tracking operations, several trials of the same action are usually required, although a single action may occasionally be successful. Rather than include several identical subtasks in the task analysis these subtasks are organized for the computer as requiring a single control action with a relatively low probability of success and a value of $(i,j)_f$ which forces the repetition of the subtask until success is achieved. On such a subtask the probability of success on any single trial may be determined for use as input data as follows: If $\bar{p}_{ij}$ is the probability of success on a single trial and p^* is the probability of at least one success after n trials, then:

$$\bar{p}_{ij} = 1 - \sqrt[n]{1 - p^*}$$

NO-STRESS RUNS

Although a principal model variable is stress, a need was encountered for simulation data from runs in which no time stress was induced. Such runs are easily accomplished by setting an arbitrarily large value for the time limits T_j so that even with an unexpectedly large number of subtask failures no stress buildup occurs. It has become a standard practice in operational run situations to make at least one such no-stress run in each series of run trials as a reference for stressed situations. These no-stress runs can also be helpful in determining reasonable T_j limits since they yield values for total time used without interaction with time stress.

EMERGENCY SITUATIONS

A dangerous situation may result from operator action or an external event during the task. In either case it is assumed that the operator on noticing the danger will abandon his task and take up the problem of survival and the sequence of operations would then change. Thus an emergency condition need not be a part of the model, per se, since the condition itself may be studied independently by establishing special emergency sequences to be simulated.

RECORDING THE RESULTS

The model is based on an iterative, interactional analysis of parameters based primarily on time and stress to produce as output actuarial predictions of the quality of performance and adjustment of men under stress. To provide a useful understandable series of presentations the model is organized so that at the completion of calculations the following four sets of recorded results may be produced:

1. Detailed results: a data-rich aggregate pertaining to individual subtasks (optional)
2. Intermediate results: a summary for each task simulation or iteration (optional)
3. Final results: a summary for all N task iterations of a run
4. Plotted results: a series of single-page graphic results at the completion of a series of runs (optional)

Selecting those items for recording and display at the various levels of output requires serious consideration during model development. It is necessary to review the types of results and their format and to select for printing those results that meet the criteria of meaningfulness to the system analyst, reasonable printing time, paper volume, and presentation efficiency.

In the optional detailed listing the following data are recorded for each subtask for each simulated operator:

1. Subtask number
2. Type of subtask
3. Essentiality indicator (N = nonessential, otherwise blank)
4. Stress and augmented stress

5. Waiting time (time spent waiting for partner)
6. Subtask execution time
7. Cumulative subtask execution time
8. Result indicator (blank = success, F = failure, I = ignore)
9. Cohesiveness index

The following information is optionally displayed in the iteration summary for each of the N iterations in a run:

1. Iteration number
2. Run number
3. Mission result indicator: overrun, task failure because of time overrun; underrun, task success
4. Total time used: the larger of the total time used by either operator
5. The following data are also provided for each operator:

(a) Operator number
(b) Stress threshold, M_j
(c) Individuality factor, F_j
(d) Time available, T_j
(e) Time used for this iteration
(f) Time overrun (time used-time available)
(g) Total waiting time (does not include item 1 below)
(h) Value of highest (peak) stress encountered
(i) The subtask number on which the peak stress occurred
(j) Stress at end of iteration
(k) Cohesiveness at end of iteration
(l) Time spent in waiting for the period on cyclic subtasks
(m) Goal aspiration difference at end of iteration
(n) Maximum difference in goal aspiration during iteration
(o) Minimum difference in goal aspiration during iteration
(p) Performance at end of mission iteration (percent subtask successes)

Figure 2.4 is a sample of the format from a run summary tabulation. It contains the following:

1. Run number
2. Total number of iterations performed, N
3. Number of successful iterations
4. Percent successful iterations
5. Time available, T_j
6. The following data are listed for each operator:

(a) Operator number
(b) Stress threshold, M_j

SUMMARY OUTPUT OF RUN   1
NUMBER OF ITERATIONS   100
NUMBER OF SUCCESSES   55
PER CENT SUCCESSES   55.0

FOR TIME DURATION 1400.00

OPR NO.	THRES HOLD	SPEED	AVAIL	USED	DIFF	WAIT	PEAK STRESS	FINAL STRESS	INITIAL GOAL	*MIN	MAX	FINAL*	LAST PERF	CYCLIC WAIT	NO. OF TASKS
										GOAL ASPIRATION DIFF.					
1	2.30	1.00	1400.0	1400.5	0.5	101.5	1.42	1.68	0.90	-0.10	0.11	0.03	0.871	0.00	202
2	2.30	1.10	1400.0	1400.5	0.5	373.6	1.44	1.68	0.90	-0.01	0.08	0.01	0.885	0.00	222

FREQUENCY DATA

TASK NO	LAST TASK COMPLETED OP 1	OP 2	TASK FAILED OP 1	OP 2	TASK IGNORED OP 1 NE	OP 2 NE
1	0	0	0	0	0	
2	0	0	1	10	0	
3	0	0	1	284	0	
4	0	0	1	5	0	
5	0	0	13	2	0	
6	0	0	22	0	0	
7	0	0	24	2	0	
8	0	0	8	3	0	
9	0	0	32	0	100	N
10	0	0	0	1	0	
179	3	0	1	0	0	
180	42	0	2	2	0	
181	55	0	1	1	99	N
182	0	0	0	7	0	
183	0	0	0	0	0	
184	0	0	0	2	0	
185	0	0	0	1	0	
186	0	0	0	9	0	
187	0	0	0	0	0	
188	0	0	0	2	0	
189	0	0	0	4	0	
190	0	0	0	0	0	
191	0	0	0	0	0	
192	0	0	0	2	0	
193	0	0	0	5	0	
194	0	0	0	3	0	
195	0	0	0	3	0	
196	0	0	0	0	0	
197	0	1	0	0	0	
198	0	0	0	1	0	
199	0	44	0	1	0	
200	0	55	0	1	0	
TOTAL	2626	2551	400	499		
AVERAGE	26.3	25.5	4.0	5.0		

TASK NO	FAILURE TIMES OP 1	OP 2	PEAK STRESS OP 1	OP 2	AVG TIME CMPLTD OP 1	OP 2	AVG START TOT STRESS OP 1	OP 2	AVG COHESIVE OP 1	OP 2
1	0.00	0.00	0	0	42.02	304.22	1.01	1.05	0.00	0.01
2	3.36	101.57	0	0	48.01	318.60	1.01	1.04	0.00	0.01
3	7.89	330.84	0	0	57.62	323.07	1.01	1.04	0.00	0.01
4	28.37	12.81	0	0	78.21	325.55	1.01	1.04	0.00	0.01
5	55.47	5.51	0	0	83.72	327.76	1.01	1.04	0.00	0.01
6	27.21	0.00	0	0	85.20	333.01	1.01	1.04	0.00	0.01
7	136.61	2.52	0	0	92.58	335.02	1.01	1.04	0.00	0.01
8	105.26	6.16	0	0	107.48	337.17	1.01	1.04	0.00	0.01
9	182.58	0.00	0	0	115.47	337.92	1.02	1.04	0.00	0.00
10	0.00	23.18	0	0	121.27	358.67	1.02	1.04	0.00	0.02
179	23.21	0.00	8	0	1383.26	1324.97	1.24	1.18	0.13	0.00
180	16.42	5.04	11	0	1393.55	1327.51	1.19	1.20	0.12	0.08
181	6.71	1.44	2	0	1407.16	1328.27	2.80	1.20	0.38	0.03
182	0.00	19.14	0	0	0.00	1331.11	0.00	1.21	0.00	0.08
183	0.00	0.00	0	8	0.00	1336.11	0.00	1.22	0.00	0.00
184	0.00	4.99	0	0	0.00	1338.49	0.00	1.22	0.00	0.08
185	0.00	4.74	0	2	0.00	1342.68	0.00	1.22	0.00	0.08
186	0.00	9.22	0	6	0.00	1343.87	0.00	1.23	0.00	0.08
187	0.00	0.00	0	0	0.00	1344.93	0.00	1.23	0.00	0.08
188	0.00	12.66	0	4	0.00	1350.75	0.00	1.23	0.00	0.08
189	0.00	28.10	0	3	0.00	1359.38	0.00	1.40	0.00	0.08
190	0.00	0.00	0	13	0.00	1361.79	0.00	1.41	0.00	0.12
191	0.00	0.00	0	2	0.00	1365.86	0.00	1.23	0.00	0.12
192	0.00	7.60	0	1	0.00	1369.12	0.00	1.23	0.00	0.12
193	0.00	0.00	0	0	0.00	1370.25	0.00	1.22	0.00	0.12
194	0.00	20.65	0	7	0.00	1376.79	0.00	1.24	0.00	0.13
195	0.00	5.93	0	2	0.00	1377.90	0.00	1.23	0.00	0.12
196	0.00	3.75	0	3	0.00	1382.73	0.00	1.24	0.00	0.13
197	0.00	0.00	0	3	0.00	1385.14	0.00	1.22	0.00	0.11
198	0.00	4.67	0	0	0.00	1389.21	0.00	1.22	0.00	0.10
199	0.00	13.57	0	14	0.00	1391.77	0.00	1.21	0.00	0.10
200	0.00	4.63	0	11	0.00	1400.20	0.00	1.25	0.00	0.08
	6894.99	6010.49								
	68.95	60.10								

6894.99 / 6010.49 TOTAL
68.95 / 60.10 AVERAGE

Figure 2.4 Run summary tabulation format.

(c) Speed factor, F_j
(d) Time available, T_j
(e) Average time used over N iterations
(f) Average time overrun (time used — time available)
(g) Average waiting time
(h) Average peak stress
(i) Average final stress
(j) Initial goal aspiration
(k) Run maximum, minimum and average difference of goal aspiration
(l) Average final performance
(m) Average cyclic waiting time
(n) Total number of subtasks simulated

7. The following frequency distributions are displayed and values presented for each subtask and for each operator. The items identified below with an asterisk are also totaled and averaged, per iteration, over all tasks:

(a) Subtask number
(b) Count for the number of times the indicated subtask was the last subtask completed before finishing the iteration or running out of time
*(c) Count of the number of subtasks failed
*(d) Count of the number of subtasks ignored
*(e) Time spent in performing subtasks that were failed
(f) Count of the number of subtasks on which the peak stress occurred
(g) Time from beginning of mission that the subtask was completed on the average
(h) Average stress prior to beginning each subtask
(i) Average cohesiveness value on each subtask

The plotted output is organized to produce one point on each of a number of optional graphs for each of up to 25 computer runs. The selectable dependent variables are:

• Average time used
• Average peak stress
• Average final stress
• Probability of task successs
• Average waiting time
• Sum of subtasks ignored
• Sum of subtasks failed

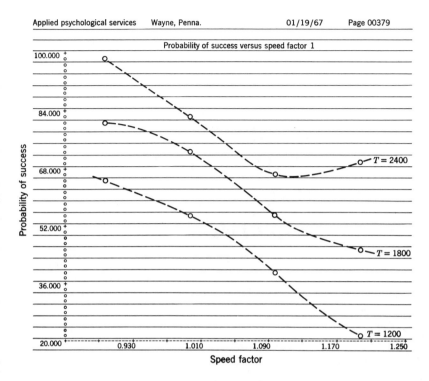

Figure 2.5 Run plot.

and the independent variables are:

- Time available
- Stress threshold
- Speed factor

Figure 2.5 is an example of a plot produced by the model following a twelve-run trial.

The model as described is the result of evolutionary development which is certain to continue as the processes of improvement by change and adaptation to various task situations continues. Nevertheless, its present form serves to illustrate its major features and the techniques employed.

III

RESULTS OF
APPLICATION OF
UNITARY OR
DUAL-OPERATOR
SIMULATION MODEL

The digital simulation model described in Chapter II has been applied to a variety of single-operator and two-operator man-machine systems. Four of these applications are described in this chapter. The first two of these applications, landing an aircraft on an aircraft carrier and launching an air-to-air missile, represent single-operator situations. The second pair of applications, simulation of an inflight refueling operation and simulation of an air intercept, represent two-operator simulations. Since the purpose of these simulations was validation of the model, the "predictions" from the model were compared with independently derived outside criterion data. The selection of specific results for presentation here has been made primarily for illustrative purposes and the four examples are treated only to this extent. Generally, in digital simulation it has become apparent that the volume of presentable results is limited only by the resources one is willing to apply. Therefore, the object of the simulation designer is to consolidate to essence, requesting and presenting only the meaningful relationships and conditions with detail available but provided only when specifically requested.

CARRIER LANDING

A single-operator mission to which an early version of the simulation model was applied was that of landing a fighter aircraft on the deck of an aircraft carrier. A task analysis indicated that from the start of the maneuver to the operator decision required in the final approach the landing task is composed of 37 basic subtasks. After subtask number 37 a decision with the following five possible first-order alternatives is made by the pilot (second-order alternatives, such as high and fast, were not considered):

1. The aircraft is approaching satisfactorily and no further operator action is required. In this case the simulation is complete.
2. The aircraft is approaching too fast and subtasks 40 through 43 must be performed.
3. The aircraft is approaching too slowly and subtasks 45 through 48 must be performed.
4. The aircraft is approaching too high and subtasks 50 through 53 must be performed.
5. The aircraft is approaching too low and subtasks 54 through 57 must be performed.

The assumption that these five listed alternatives were equally likely was implemented by the technique of decision subtasks. Each subtask of the landing sequence was required to be repeated until it was performed correctly.

A total of 210 sec is available to the pilot during which the sequence of subtasks is to be performed; that is, $T_j = 210$ for all runs of the landing task.

In order to obtain some measure of the agreement between the results obtained from the model and the actual performance of pilots outside criterion data were obtained. The outside criterion data derived from ratings of Landing Signal Officers for 162 carrier landings are presented in Table 3.1.

As may be expected values of M_j and F_j were found to have a pronounced effect on the number of subtasks that were completed by

Table 3.1   Outside Criteria Data for the Carrier Landing Task

Class of Landing	Flight Description	Frequency	Percentage
1	Poor landing, or given waveoff	41	25.3
2	Mediocre landing, improvement highly desirable	59	36.5
3	Good approach but improvement possible	41	25.3
4	Excellent landing	21	12.9
		162	100.0

the simulated pilot. In the worst case encountered the task was failed by the operator after completion of as few as 27 subtasks. In the best case no landing failures were encountered for an entire 100-iteration run. The data were grouped into classes of landings on the basis of the last subtask performed successfully to facilitate summarization and comparison with the outside criterion data. Class 1 was defined to include those landings that were not completed to the decision point following subtask 37. Class 2 included those simulations in which the decision point was reached, further adjustments were required but these were not completed successfully before all allotted time was expended. Class 3 included those simulations in which the additional adjustments were required and completed satisfactorily. Class 4 included those simulations in which the decision point was reached successfully, no further adjustments were required, and the last subtask completed was 37. Thus class 4 corresponds to a perfect landing. Class 3 corresponds to a landing in which the pilot completed his job but required a number of last second adjustments.

These data in the four classes were further aggregated into two divisions—those unsuccessful simulations in which the task was not completed (classes 1 and 2) and those successful simulations in which it was (classes 3 and 4). The failure grouping, plotted in Figure 3.1 as a function of the stress threshold, yields a value for the number of failures in each of the computer runs.

If one accepts the assumptions that the aggregate of classes 3 and 4 as given in Table 3.1 (good and excellent landings) corresponds to a simulation in which the pilot has completed all subtasks required (that is, they are equivalent to classes 3 and 4 of the computer simulation) and that the aggregate of classes 1 and 2 as given in Table 3.1 (mediocre and poor landings) corresponds to the classes 1 and 2 of the computer simulations, then the probability of failure from outside criteria data obtained from Table 3.1 as 0.618 (50 out of 81 failures) may be compared to computed data. This value of 50 is indicated on Figure 3.1 as an arrow at the Y axis at $M_j = 2.3$.

Since a primary aim of this study was to evaluate the model and determine conditions under which it agrees with actual events, this value of $M_j = 2.3$ is a value of a basic parameter under which agreement is obtained between computed results and outside criteria. Realizing, however, that the individual points in Figure 3.1 are indicative only within certain confidence limits, the M_j value should more accurately be given by a range of values. This range of values taken from Figure 3.1 is approximately $1.95 \leq M_j \leq 2.80$ corresponding to failure frequen-

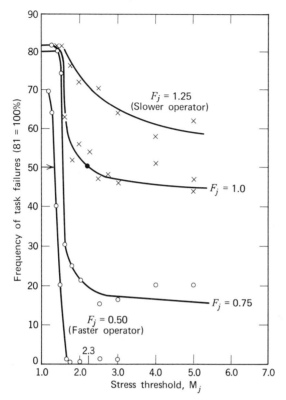

Figure 3.1   Frequency of task failures, landing task.

cies from 46.5 to 53.5 out of 81 or failure probabilities of from 0.575 to 0.66.

Chi-square tests for the significances of the differences between the outside criterion data and the summaries of computed data indicated that a wide range of nonsignificant differences was achieved. Moreover the entire area of M_j values ($1.95 - 2.80$) which were indicated as the most possible valid model parameters yielded nonsignificant differences when compared with the outside criterion.

The results of the simulations indicated that in every simulated landing some "idle time" was encountered by the pilot on the "downwind leg" prior to "picking up the mirror." These data were verified against pilot opinion which suggested that on making the approach the pilot is not overloaded on the "downwind leg" but that he is very busy during the "final approach."

MISSILE LAUNCHING

Another simulated one-man task was that of the launching of an air-to-air missile. The task analysis for the launching task indicated 22, 23, or 24 basic subtasks from the start of the maneuver to the final "break away." Subtask 13, a decision subtask, allowed the computer to ignore subtask 14 during simulation 25 percent of the time. Following subtask 19 a decision is made in accordance with the following three equally probable alternatives:

1. The target is properly centered on the radar scope, therefore subtasks 21 through 24 are to be performed.
2. The target is too high on the radar scope, therefore subtasks 26 through 30 must be performed.
3. The target is too low on the radar scope, therefore subtasks 31 through 36 must be performed.

Because of the nature of the launching task, subtask number 18 could not be initiated until "lock on" had occurred. This was allowed after a minimum of 79.2 sec following the start of the maneuver.

The value of T_j was determined from the best estimate of the distance traveled by the intercept aircraft, its average speed, and a break-away distance of 2 miles between the attacker and his target to have a value of 184.8 sec; if a 4-mile break-away distance is employed, the time limit becomes 171.6 sec. A launching was considered to be successfully completed only if all subtasks, including break-away, were completed prior to T_j elapsed sec. In practice if insufficient time remained to complete the required action prior to missile launch, the pilot would, of course, give up the launching sequence and break-away.

As in the landing task the parameter values were found to have a pronounced effect on the number of subtasks that were completed during the simulation runs. In the worst cases encountered the task was failed by the simulated operator after completion of as few as eight subtasks and, in the best case, no launching failures were encountered for an entire run. A summary of the results concerning task completion is presented in graphical form as Figure 3.2 for all runs.

Outside criterion data for the launching task obtained from fleet records indicated that 16 out of 88 attempted launchings in the real situation resulted in failure. This figure is indicated in Figure 3.2 as an arrow on the ordinate. Inspection of Figure 3.2 indicates that the $F_j = 1.00$ curves do not intersect the outside criterion at any point. On the other hand the $F_j = 0.90$ curve crosses reality and underpredicts the actual number of launching failures at M_j values between 1.95 and 2.80.

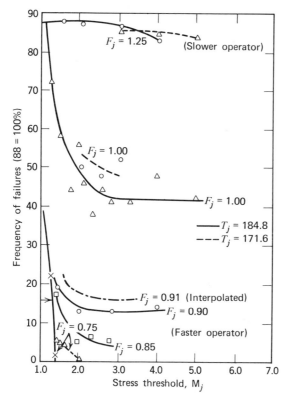

Figure 3.2   Frequency of task failures, launching task.

An interpolated curve of $F = 0.91$ (dotted curve, Figure 3.2) would coincide with the outside criterion data with M_j values in the specified range.

DISCUSSION OF LAUNCHING AND LANDING

These results may be interpreted as indicating reasonable concordance between the model's prediction and outside criterion data for the two applications discussed. As stated above it is noted that data from Figure 3.2 would, in fact, cross the outside criterion at a value for M_j, which lies within the range of values predicted from the landing mission. Thus agreement would accrue if the missile launching squadron was faster than average and the landing squadron was slower than average by a combined total of 9 percent. It is also known that the outside criterion data for the missile launching task were obtained during daylight missile test situations in which the attack was staged; the carrier landing data were obtained under more varied and realistic operational conditions. The favorable missile test flight conditions may account for the comparatively low failure rates shown by the pilots involved in the launching analysis.

Comparison of Figures 3.1 and 3.2 indicates that

1. The model yields relatively smooth curves whose form is characteristic and repeatable from one task to another.

2. The data serve to confirm the gross expected results that higher success probabilities are obtained by faster operators and by operators who can tolerate more stress without breaking, or by an increase in the time limit.

3. For the average pilot in the launching task a failure probability between 0.40 and 0.435 was indicated in the stress threshold range coincident with that previously derived for landing aboard a carrier.

INFLIGHT REFUELING

The objective in this task is the midair insertion, by the pilot of a strike or refueling aircraft, of a probe into a drogue extended by the pilot of a tanker aircraft. In order to determine the sequence of subtasks involved together with the relevant auxiliary data four experienced pilots were asked to list the subtasks sequentially. These four independent analyses were synthesized into a composite best estimate and resubmitted to the pilots for review. The analysis was then revised in accordance with their suggestions and resubmitted for further review. At this point the pilots unanimously agreed on the procedures outlined and the sequence was therefore adopted. The various input data were determined from the basic task analysis and the total times for each member of the team were determined on the basis of estimated proficiency for the "average" pilot.

Figure 3.3 presents schematically the time relationships of performance of the two operators using average execution times, average waiting times, and assuming no subtask failures. Operator time limits and decision probabilities are also shown. A time diagram of this type is found helpful as well as informative and is recommended prior to simulating each task using this type of digital simulation.

At the start of the task the tanker and the strike aircraft are assumed to be flying abreast of each other at a rate greater than optimal for inflight refueling. The pilot of the tanker aircraft is assumed to be aware that refueling is to take place but not to have prepared for it. Initiation of the task takes place when the pilot of the strike aircraft reports,

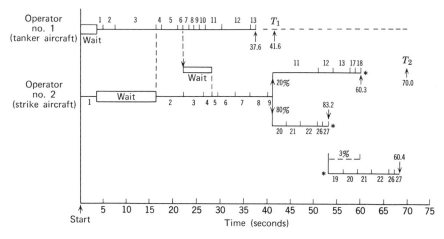

Figure 3.3 Schematic time diagram of the inflight refueling subtasks.

"Ready to refuel." The tanker pilot acknowledges by reporting "ready" to extend drogue and reduces airspeed while the strike pilot waits.

A wait is again shown for the strike aircraft at subtask 5 because the strike pilot cannot observe the fully extended drogue until the tanker pilot has completed his subtask 6 (extending the drogue). On seeing the extended drogue the strike pilot moves his aircraft astern of the tanker.

Subtask 10 is a decision subtask for the strike pilot and branching occurs. About 80 percent of the time the computer considers subtask 20 next; 20 percent of the time the computer considers subtask 11 next. The purpose of this branching is to provide for two different amounts of time being spent on "observing steadiness of drogue" ($i = 11, 20$) as might be required with varying air turbulence. Except for this difference between their initial subtasks, both branches are identical. If the decision is made to follow the branch headed by subtask 11 and if on the last subtask of that branch ($i = 18$) the probe is not successfully inserted, the computer takes the other branch on the next attempt. It is assumed in this case that the previous observations of drogue motion will be recalled and applied. On failure to execute the last subtask of the alternative branch that same branch is repeated until success is finally achieved.

Forty-four computer runs, each consisting of 100 iterations, were performed for the refueling task. A summary of the primary results of the runs is shown in Table 3.2, which gives the total number of

Table 3.2   Percentage Successes: Refueling Task*

Stress Threshold of Pilot of Strike Aircraft (M_2)	Stress Threshold of Pilot of Tanker Aircraft (M_1)			
	1.25	1.50	3.00	4.00
1.25		86:74:40:10		
1.50	91:65:47:9	92:68:55:10	90:75:54: 6	91:73:52:11
3.00		90:81:56:14	89:74:46: 9	89:79:51: 9
4.00		92:81:49:16	94:79:49:14	90:77:53:10

* The four figures in each set are for relative operator speeds of $F_1 = F_2 = 0.9$: 1.0: 1.1: 1.3, respectively.

successful simulated refuelings taking the F_j values for the two pilots as equal $(F_1 = F_2)$ at values of 0.9, 1.0, 1.1, and 1.3.

Table 3.2 shows that faster simulated pilots achieved very substantially greater success than slower. This relationship is shown in Figure 3.4. Here average percentages of success over the 11 sets of runs are plotted for each of the 4 F_j values. The shaded area shows the range of values obtained in the runs.

In order to find out how well the results obtained by the model agree with reality data were collected about the degree of success achieved by operators of known capability. Of 16 test flights 10 resulted in success-ful engagement of the probe on the first attempt. The probability of task success for these pilots, 62.5 percent successes, is shown on Figure 3.4 as a solid arrow on the ordinate. The model produced a similar probability of success when $F_1 = F_2 = 1.05$; that is, for operators who are slightly slower than the average. This value is, however, well within the one-sigma confidence limit that may be placed on the outside cri-terion value as shown by the dotted arrows on the ordinate of Figure 3.4.

The effect of M_j on percentage success was small for the tanker pilot as would be expected from the comparatively easy role he plays. Effects for the strike pilot were somewhat greater, although probability of suc-cess was relatively stable as M_2 increased above 2.0 to 2.5. Below these values of M_2 probability of success tended to decrease, although the changes were small compared with those resulting from changes in the value of F_j.

Although the model indicated no need for idle time for either operator, waiting time was encountered for both team members. No significant effect could be attributed to variation in M_j values but the average wait-ing time for runs over all M_j values suggested that the effects of F_j on

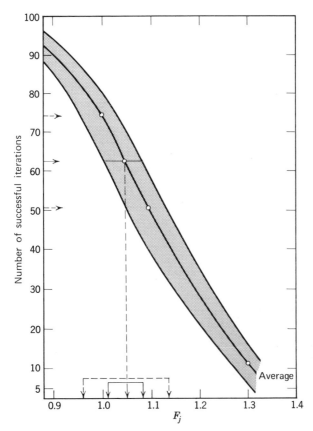

Figure 3.4 Percentage of task successes as a function of operator speed (F_j)—refueling task.

waiting time were approximately linear. As one might expect faster teams spent less time waiting for each other. The tanker pilot spent about 0.3 sec more and the strike pilot about 1.3 sec more in waiting for every 10 percent decrease from average in team speed.

The average time remaining for the strike pilot (who was the pacemaker of the team) after successful completions of the task indicated little overloading for the average and slightly faster than average pilot. However, for the somewhat slower than average pilot $(F_j = 1.1, F_j = 1.3)$ the time allowed for the task was tight and some overloading was present.

As expected both the peak and the terminal stress values increased with increasing F_j values, indicating that slower operators tend to build up greater stress than faster ones and the strike pilot built up greater stress than did the tanker pilot. Review of the data from individual runs indicated that the peak stress condition occurred near the end of the runs—for example, when the strike pilot closed in for actual probe insertion.

The model calls for either operator to skip each nonessential subtask whenever his stress exceeds unity. As an example the number of nonessential subtasks ignored by the strike pilot over all M_2 values are presented as Figure 3.5. Contrary to expectation this suggests that operators with lower stress tolerance skipped fewer nonessential subtasks. In conformity with expectation, however, slower operators skipped more nonessential subtasks.

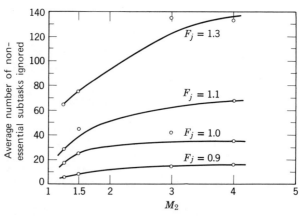

Figure 3.5 Average number of nonessential subtasks skipped per run by simulated strike pilots—refueling task.

The team cohesiveness data are presented in Figure 3.6 as a function of varying stress thresholds for the strike pilot. This figure indicates, as expected, that faster teams were more cohesive, that higher stress thresholds yielded better teamwork, and that a level of C_{ij} equal to unity occurred when $M_2 = 1.25$, $F_j = 1.0$, and for $M_2 = 2.25$, $F_j = 1.1$.

AIR INTERCEPT

In this task the actions of a radar observer and a pilot are simulated in the intercept if an intruding aircraft by an advanced, supersonic naval aircraft. A summary flow diagram of the intercept run simulated, a team attack, is presented as Figure 3.7.

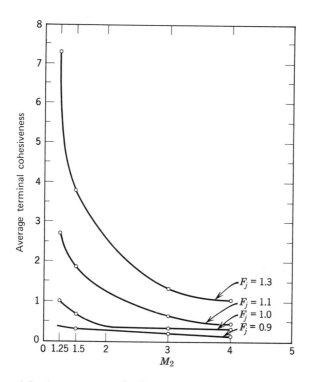

Figure 3.6 Average terminal cohesiveness as a function of M_2 and F_j.

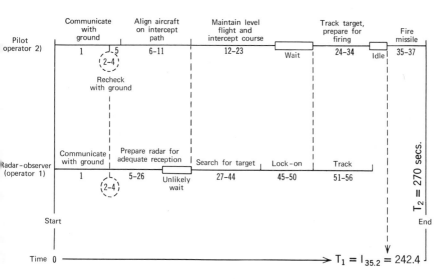

Figure 3.7 Summary time diagram of the intercept task (not to scale).

The task is initiated with the simultaneous reception by the pilot and the radar observer of the target vector from ground control. From here until he completes a turn that will align his aircraft with the preliminary intercept course the main responsibility rests with the pilot. Following the completion of subtask 11 (report to ground control) by the pilot, the main responsibility shifts to the radar observer and continues with him until he has successfully completed his subtask 50. During subtasks 11 to 24 the pilot's concern is solely with maintaining level flight on a fixed vector. On completion of subtask 50 by the radar observer the pilot checks the steering dot position and assumes the main responsibility until the end of the intercept. The total situation essentially allows an idling time before which the pilot cannot accomplish his subtask 35.

If the radar observer successfully completes subtask 1, he continues with subtask 5 without waiting for any action by or response from the pilot. On the other hand should the radar observer fail at subtask 1, he must wait until the pilot has successfully executed his fifth subtask before communicating with him. A second waiting time may occur for the radar observer on subtask 27, which he cannot accomplish until the pilot has completed the turn that positions the aircraft on the intercept path. Before this point the radar observer's concern is with the preparation of the radar for the most adequate reception. In practice and in simulation, however, this wait is very seldom required.

The values of the idling time and the time limits T_1 and T_2 were calculated from assumed relative positions and speed of the interceptor and target aircraft.

Success for the intercept task is defined within the model as the completion by the pilot within the time allowed of all 37 subtasks and the completion by the radar observer of at least the 46 subtasks that were classified as essential.

Thirteen computer runs, each consisting of 75 simulations, were performed. Table 3.3 shows for operators of various speeds and stress thresholds the total number of successful intercepts in each run. The table suggests that M_j had little effect within the range of values investigated but that as expected faster teams achieved substantially more successes than slower teams. The mean number of successes by F_j values over all runs is given in the lower curve of Figure 3.8. The shaded area shows the range of values obtained in the runs. The leveling of this curve at its left ($F_j = 0.95$) is noticeably steeper than that obtained in the refueling simulation. This is believed to result from the fact that in the intercept task each operator's probablity of success in his first subtask was 0.6. If both the pilot and radar observer failed at their respective first

Table 3.3   Number of Successes out of 75: Intercept Task as a
Function of Stress Threshold°

Stress Threshold of Pilot (M_2)	Stress Threshold of Radar Observer (M_1)		
	1.5	2.0	3.0
1.5	62:63:25	67:61:30	
2.0	65:61:31	67:63:26	
3.0			-:60:-

* The three figures in each set are for relative operator speeds of $F_1 = F_2$
= 0.95: 1.0: 1.1, respectively.

subtasks, an additional 14.1 sec for the radar observer and 13.4 sec for
the pilot would, on the average, have been required. These delays cannot
run concurrently, so that when this dual failure occurs as it did 16 percent of the time, the team effectively starts with a 27.5 sec "handicap."

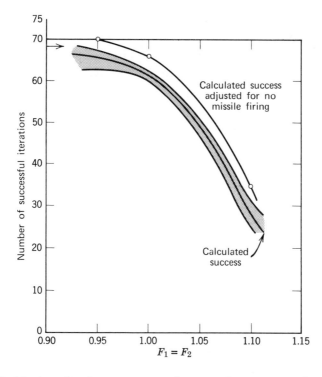

Figure 3.8   Number of task successes as a function of operator speed (F)—intercept task.

Data were obtained on actual intercept success for a beam attack with intercept and intruder aircraft of the types simulated. Of 25 test runs two were "unsuccessful" and the remainder "successful," yielding an approximation of the probability of success for the two-man team of 92 percent, reflected as an arrow on the ordinate of Figure 3.8. In these field tests, however, the missile carried by the aircraft was not actually fired; therefore the pilot completed only 36 subtasks and an adjustment was made for the average firing time of 1.1 sec. The number of successes plus those simulations in which failures occurred because of less than 1.1 sec overrun are shown as the upper curve in Figure 3.7. The difference between the value shown on this curve for $F_j = 1$ and that of the field trial is 4.8 percent, which is within the 2σ (± 7.5 percent) confidence limits calculated from the simulated results.

Waiting time occurred during the simulations for both team members. Average waiting time data in relation to stress thresholds showed that the pilot spent more time waiting for the radar observer than vice versa. As expected faster teams tended to show longer waiting times and to: complete successful runs with more time on hand, build up less stress, omit fewer nonessential subtasks, and show higher cohesiveness.

The pilot tended to build up greater stress than did the radar operator and while the stress threshold exerted little effect on task success, it caused slower pilots to skip more nonessential subtasks in order to achieve success.

CONCLUSIONS

It may be concluded that in these validatory studies the model yielded results that largely appear to conform with reasonable expectation and with criterion data. While the technique is certainly fallible and dependent on a rational but synthetic logic, it appears that for predicting operator effectiveness on tasks similar to those simulated the model may be used with some degree of confidence.

A summary of mission simulation results, which have been compared with actual performance of the same missions, is shown in Figure 3.9. This figure indicates by the arrow the probability of success derived from operational outside criterion data for each mission. Correspondingly the ranges of success probabilities predicted by the computer for the same missions are shown in the respective vertical bars. Various values of the operator individuality factor F_j are shown. A value that exceeds 1.0 indicates a slower (or less proficient) operator. It has been concluded from these past efforts that with adequate care in developing mission sequences and assigning numerical values for the required computer

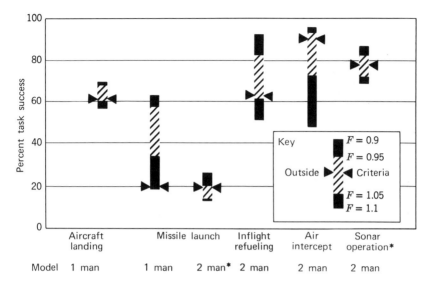

Figure 3.9   Results of validations of digital simulation model.

input data the model's predictions will be in reasonable accord with outside criterion data, normally erring on the conservative side. The technique has been found particularly useful in comparative runs in which the effects of parameter variation are determined. When the findings resulting from the tasks described are coupled with others (including results obtained by comparing laboratory task performance with simulation results) considerable support is given to the contention favoring validity of the model.

The method of simulating one or two-operator man-machine systems seems applicable to a variety of situations. It appears to be practical and to yield results which will help in making decisions regarding the potential of a given man-machine combination. The model yields results that are internally consistent and variation of parameter values yields reasonable effects on the predicted system operation. Thus the utility of simulating certain psychological and man-machine interactive processes is demonstrated.

The next chapter demonstrates how a logic similar to that described in the present chapter can be modeled when larger groups and social psychological as well as individual psychological variables are involved.

IV

GROUP SIMULATION— QUALITATIVE AND CONCEPTUAL CONSIDERATIONS

SCOPE

This chapter presents introductory and psychological background material relating to the second digital man-machine simulation model. Since this model is concerned with group performance, the chapter is principally concerned with various aspects of the logic underlying the psychosocial group-oriented variables that are believed salient and their influence on group behavior, which make them appropriate for digital simulation. The specific application of these aspects—that is, the method by which these variables are reflected and interact in the mathematics and logic of the model—is discussed in Chapter V. Additionally, for purposes of continuity certain aspects of the psychological representation are discussed in Chapter V.

OVERVIEW OF MODEL

As a framework for the discussion we conceive of a group of supervisors (naval officers) and workers (enlisted men) who form a "crew" and who man a system for a given period of time. The crew, characterized by numerical values selected by the computer, "performs" certain assigned work (for example, system operation and maintenance jobs) which is broken down into subtasks called "action units." In order to accomplish each action unit men are "selected" by the computer from the crew to form a group. Thus the group in this model is conceived as an ad hoc formation of men assigned to perform a specific action unit. The group may or may not have worked together as a team previously. The total of all action units over a number of days constitutes a mission.

The model applies primarily to systems maintained, or operated, or both, over a relatively long period (days or weeks) by crews of up to several dozens of persons. The situation implies that the crew composition does not change during a mission except for incapacitation because of sickness.

Since much of the simulation process is devoted to formation and simulation of small groups of men from the crew, the major effects of the simulation are not realized for small work forces, say of less than 10 to 15 men.

Whereas the two-man model was concerned with stress and other variables relating to relatively short missions, this model is used to predict the qualities of larger man-machine systems such as:

1. System efficiency as a function of crew size and mission time
2. Crew morale and cohesiveness during the course of the mission
3. Time devoted to equipment repairs
4. Manpower time shortage by type of personnel as a function of crew size
5. Proficiency of crew members
6. Man-hour loadings and overtime

Although the aspects of organizational doctrine, personnel policies, replacement limitations, and medical factors are not ignored in the model, the psychological aspects have been given prime emphasis in affecting performance. The digital simulation does not attempt to duplicate, in a computer, every minute aspect of an actual group-interactive situation. Rather the technique predicts a real life criterion by a number of factors believed to capture the essence of those variables that current leaders in the field of social psychology seem to agree on as affecting the performance of the human who is a part of a closed social group.

We begin with an interpretation of those factors believed salient by others, set these factors in a form that will allow digital computer simulation, and then manipulate the factors in order to predict the criterion situation. Accordingly the model concepts are largely based on current references and theoretical texts with respect to delineation of group process and social unit effectiveness variables. Therefore the review and summary of those variables, as included in the model conceptualization, are the functions of the balance of Chapter IV. The reader concerned primarily with the model's features, as opposed to its theoretical basis, may turn directly to Chapter V without depending on the material in the balance of this chapter.

MAJOR VARIABLES IN SMALL-GROUP THEORY

Hare (1962) divided the small-group field into three gross areas: (1) group processes and structure, (2) interaction process variables, and (3) performance characteristics. Under the first classification he included interaction elements, norms and social control, interaction and decision processes, roles and interpersonal choice. Under interaction process variables he included personality, social characteristics, group size, task communication network, and leadership. Finally, in the performance characteristics classification he included the interactions of the individual with the group and the group with other groups.

Cartwright and Zander (1962) included much of the same information in their discussion but spoke of group cohesiveness, group pressures and standards, individual and group goals, leadership and group performance, and structural properties of groups. One difference between Cartwright and Zander and Hare seems to be in terms of the separate classification of group cohesiveness.

Gilchrist (1959) in discussing the areas of greatest amount of work classified variables in terms of social influence and opinion and attitude change, social perception and impression formation, authoritarianism and acquiescence, social interactions and group processes, and cross-cultural studies.

Roseborough (1953) in an attempt to ". . . evaluate the present state of empirical knowledge in the field of small group study. . ." classified research results in terms of (1) social structure: diffuse authority (attitude change) and specific authority (leadership); (2) cultural variables: group affiliation and experimental instructions; (3) situational variables: task problem, size of group, spatial position, and communication patterns; and (4) personality variables: leadership, social interaction patterns, and sociometric choice. This classification scheme seems to parallel Hare's classifications, as well as Cartwright and Zander directly; many of the variables listed by Gilchrist seem to be subsumed by Roseborough under social structure and personality variables.

Sells (1961a) and McGrath (1962) have paralleled these previous classifications more or less directly. Sells added, however, the reinforcement provided by the group (potency of group to members). Sells (1961b) also included environmental stress, such as isolation and deprivation as important variables. Reinforcement by way of social power and stress were also included by George (1962), in addition to the variables listed by Hare, Cartwright and Zander, and Roseborough. Finally, Altman and Terauds (1960) have provided a basic reference that specifies major and minor variables in the small group field as well as their interaction.

SUMMARY OF SMALL-GROUP VARIABLES

On the basis of the classification systems found in the literature Table 4.1 summarizes the major small-group variables suggested by these authors, the consensus of opinion, and gives a general reductive panorama of the important small-group variables. The names of corresponding variables and features of the group crew model are also shown in Table 4.1, although the discussion of their function, interaction, and utility is postponed for Chapter V.

A single example of this general classification might be provided by drawing on the analysis of the dyad and its application to larger groups. According to Thibaut and Kelley (1959) a relationship is developed and maintained by way of cost-reward considerations: Reward is *reinforcement* and at the extreme, cost is stress. The ratio of reward to cost might become a reflection of *cohesiveness*. Individuals and groups develop *norms* and *roles, status* and *goals* by way of organization in terms of providing high rewards at low costs. Power is associated with roles and also in terms of control or reinforcement for the attainment of individual and group goals. Further, the size of the group may determine to some extent the *communication* patterns in terms of providing some reward at minimum cost. *Task performance* is based essentially on the same considerations developed for person-to-person relationships. Finally, *personality* and *social-cultural characteristics* are important in determining what kinds as well as the values of the rewards and costs incurred in any relationship.

The succeeding sections of this chapter present a discussion of each of the concepts identified in Table 4.1 in relation to the group simulation model.

NORMS AND GOALS—ORIENTATION

For the present model the concepts of norms and goals have been subsumed under a rubric called "orientation." The term "orientation" is taken to mean primarily attitude and values that combine to form the context for a motivational set. The motivational set in part determines

Table 4.1   Major Small Group Variables and Author Agreement

Variable*	Author Agreement	Related Model Variables
Norms, group and individual	Hare, Cartwright and Zander, Sells, Altman, and Terauds	Average performance time,† operator proficiency‡
Goals, group and individual	Cartwright and Zander, Sells, Altman and Terauds	Self, crew, and mission orientations‡
Cohesiveness, solidarity	Roseborough, Cartwright and Zander, Sells, Altman and Terauds	Morale‡, cohesiveness indices‡
Pressure, reinforcement	Sells, Altman and Terauds	Psychosocial efficiency‡, performance deviation from expectation‡
Communication network	Hare, Roseborough, Altman and Terauds	Communications efficiency‡, station identification†
Environmental stress	George, Sells, Altman and Terauds	Emergency probability†, manpower shortages and conflicts‡, repairs‡
Leadership	Hare, Roseborough, Cartwright and Zander, George Sells, Altman and Terauds	Personnel types†, pay level‡, promotion‡
Roles and status	Hare, Sells, Altman and Terauds	Crosstrained personnel‡, training tasks†
Group size	Roseborough, Sells, Altman and Terauds	Group size and selection criteria‡,
Task performance	Hare, Roseborough, Sells, Altman and Terauds	Proficiency‡, crew efficiency‡, task essentiality†
Personality	Hare, Roseborough, Gilchrist, Sells, Altman and Terauds	Not simulated per se
Social-cultural characteristics	Hare, Roseborough, Gilchrist, Altman and Terauds	Crosstraining†, sickness‡, length of workday†, work shifts†, crew size‡

* This column is composed of modifications of author classifications in some cases
† Supplied as input to computer
‡ Calculated by computer

perception of the situation and possesses implications both for and against isomorphic representation of the situation. Thibaut and Kelley (1959) presented a similar notion as follows:

". . . (a) person maintains a more or less constant orientation or intention . . . (set) . . . . The specific set or sets aroused at any given time depends upon instigations, both from within the person (e.g., need or drive states) and from outside (incentives, problem situations, or tasks confronting him, experimental instructions), and the reinforcement previously associated with enactment of the set. The stability of the set depends upon the temporal persistence of the stimuli that serve to investigate it."

Newcomb (1959) also provided a similar notion in reference to the *organization* of psychological processes. To Newcomb, thinking, feeling, perceiving, and performing are the basis for an orientation and this orientation represents a "predisposition" to respond in a given direction.

Orientation, as defined within the model, applies separately to the individual, the work group, and the total crew. Orientation in relation to an action unit is in terms of the perception of the performer(s) of derived benefit; that is, will the individual, group, crew, or mission primarily be supported by action unit completion. This conception allows for a negative or aversive, as well as a positive or incentive, function for a given action unit. Thus within the model the orientation of the person, the work group, and the crew as well as the type of action unit affect performance efficiency which is a function of the joint interaction of these components. The basic notion of action unit orientation and its relationship to the other orientations follows from Thibaut and Kelley's (1959) suggestion that ". . . person-task phenomena can be analyzed in the same terms as person-person interactions," and that an extension of reward-cost considerations suggests a way of classifying tasks.

ORIENTATION OF THE INDIVIDUAL

It is assumed that for a given action unit an individual's orientation may be primarily directed toward himself, the crew, or the mission. A high value for self or individual orientation means that the individual crew member is primarily egocentric and functions best when personal benefits are high. Crew membership and action unit completion is perceived by this individual as secondary to individual benefit and therefore his dependence on the group is low. The performance efficiency of a self-oriented individual will be highest when the work is such that it satisfies egocentric needs.

High crew orientation of an individual indicates that he is primarily altruistic and functions best when crew benefits are high. This individual places crew goals above personal or mission completion goals. Crew membership is primary and individual benefit secondary. The highly crew-oriented person's dominating motivation is directed toward the maintenance of the crew and consequently the individual as a member. Cartwright and Zander (1962) suggest this as "Group membership, which was only instrumental at first has become an end in itself." The major implication associated with a crew-oriented individual is that his performance efficiency will be highest when the action unit is such that its successful completion will support crew integrity.

A high mission orientation for an individual means that he is directed primarily toward action unit or task completion and only secondarily toward individual or total crew needs. This individual is motivated toward the job for its own sake; in terms of performance efficiency his orientation will have a positive effect when the action unit is such that its successful performance will materially contribute to mission success.

The concepts of individual needs, values, goals, and norms are related to an individual's orientation in that his orientation is assumed to be the psychological integral of these concepts. This integral is thought of as possessing a motivational value rather than as a personality index. As such an individual's orientation is related to performance efficiency in a more straightforward manner.

ORIENTATIONAL MIX AND INTERDEPENDENCE

Except in extreme cases an individual is not considered to be completely oriented in any one direction and to be without some loading in the other orientation elements. This may be represented by the example below in which values represent the strength of the orientations for three individuals:

		Orientation		
Individual	*Self*	*Crew*	*Mission*	*Sum*
A	0.3	0.1	0.6	1.0
B	0.2	0.4	0.4	1.0
C	0.8	0.1	0.1	1.0

Individual A would be characterized as being primarily mission oriented since his loading on this orientation is highest; however, he is not completely mission oriented since he possesses weightings on the other orientations. Individual B is characterized as mildly crew and mission oriented. Individual C is characterized as being strongly self-oriented. As the above representations imply the orientations are not independent; that is, high values in one orientation are associated with low values in the other and the sum of orientational strengths must equal unity.

CREW ORIENTATION

As in individual orientation the crew's orientation may be primarily directed toward the individual, the crew, or the mission. Individual or self-orientation for the crew means that the crew is composed of individuals who are primarily self-oriented and therefore are directed toward the satisfaction of egocentric needs. Since the individuals composing the crew vary in their egocentric needs, norms and goals are not well defined and as a consequence crew solidarity is low.

A crew-oriented crew is one that is composed of individuals who are primarily directed toward the maintenance of the crew. Here, on the basis of the perceived similarity of values, it would be expected that norms and goals would be well defined (Stotland, Cottrell, and Laing, 1960) and as a consequence group solidarity and cohesiveness high.

The crew that is primarily mission oriented is one that is directed toward action unit completion and only secondarily toward crew or individual needs. Their primary goal is mission completion. The concepts of crew needs, goals, and norms are related to crew orientation in that the dominating motivational set(s) summed over the individuals reflects group needs, which in turn determine goals and norms.

The pattern of the strength of the various orientations defines the crew with the assumption that in general no orientation will be pure—that is, without weightings on the others as well—and the sum of the three crew orientational values again equals unity.

ACTION UNIT ORIENTATION

As the above description implies the action unit being performed by an individual or group is considered to be perceived in terms of derived benefit from action unit completion and may assume the characteristics of a goal. Satisfactory completion of the action unit may then be assumed to represent goal attainment and to constitute a reinforcing state of affairs. In order to provide an orientation for an action unit the perception of derived benefit in terms of needs and need satisfaction becomes the critical factor. Therefore an action unit that is individually oriented is conceived in the present model as one that, when completed, satisfies egocentric needs; crew orientation for an action unit is conceived in terms of the degree to which action unit completion serves to maintain the crew as an entity; mission orientation reflects the degree to which mission completion is served by action unit completion. Finally, as in the other orientations no action unit would be expected to be unilaterally oriented. Unlike the others, however, action unit orientations are independent; that is, they may support the individual, or the crew, or the mission, or any combination of these; their sum is not fixed.

COHESIVENESS

In accordance with the interpretation given by Hare (1962) a distinction is made between morale and cohesiveness. Within the model morale is represented as a predisposition to accomplish mission goals. Cohesiveness is represented in terms of homogeneity of attitudes, values, and goals (orientations). This conception appears to be in accordance with the thinking of workers such as Stotland and his associates (1960) who related cohesiveness to perceived similarity of attitudes, values, and goals. Thus cohesiveness is defined within the model as a function of the variance of the crew's orientational component of highest intensity and of the crew's orientational component of highest variance.

PRESSURE AND REINFORCEMENT

In order to provide some framework from which to view the relationship between pressure and performance efficiency, the individual's, the work group's, and the crew's orientations again assume prime importance. Pressure implies a force caused by an imbalance. Within the present model possible conditions of imbalance are based on the orientations of the group, the crew, and the action unit. An analysis is now presented of conditions for five distinct situations called "cases," which resulted from the ordering and grouping of the 27 combinations of dominant orientation values. This categorization is useful in the development of the basic relationships used in the model.

Case A represents a situation in which the group and crew are oriented in the same direction (i.e., the orientational components of highest value are the same) and the action unit orientation supports both. Here no deviant pressure is exerted and in addition successful completion of the task reinforces both the group and the crew. Optimal conditions prevail and performance efficiency is expected to be high.

Case B represents the situation in which the group and the action unit are oriented in the same direction and the crew is oriented in a different direction. Here the crew will exert pressure on the group to conform but since the action unit's orientation supports the group, the pressure will be rebuffed. Since the orientation of the group performing the action unit is supported by the orientation of the action unit, while performance efficiency will be somewhat lower than in case A it will still be relatively high. Successful completion of the action unit reinforces the orientation of the work.

Case C represents a situation in which the crew and action unit are oriented in the same direction but the group is oriented in another direction. Under these conditions in addition to being oriented in a different direction the crew is supported by the action unit's orientation; hence even more pressure will be exerted on the work group to conform than in case B and as a result the group will tend toward conformity. Since the group conforms to crew pressure, performance efficiency, while lower than in cases A and B, will be moderately high. Successful completion of the action unit reinforces the crew orientation and in addition the conforming behavior of the work group.

Case D represents the situation in which all components are different. Under these conditions the crew will exert pressure on the group to conform to crew values. Since the action unit supports neither the group nor the crew, however, it is not known whether conformity to crew values or to the orientation of the action unit will develop. If conformity to crew values develops, the situation approximates that described below for case E; if task orientational leanings develop, the situation resembles case B. It is assumed that case E results in low performance efficiency, whereas case B results in fairly high performance efficiency. In the long run the net result is a moderate performance efficiency value since on any one occasion the performance efficiency value may assume any value from low to high depending on the frame of reference adopted. The value for performance efficiency under these conditions is assumed to be lower than in cases A, B, and C. Thus in Case D performance facilitation is relatively low.

Case E represents the situation in which the group and crew are oriented in the same direction but the action unit is oriented in a different direction. Under these conditions the crew supports the work group's orientation, which is in a direction other than that of the action unit. Further, since the action unit supports neither the group nor the crew, facilitation will be lower than for any of the other conditions.

The model assumes that an individual's orientations are dynamic and that experience in a context will affect an individual's points of view. Accordingly an individual's orientations are modified as a result of his experience following each action unit on which he performs. When an action unit is successfully performed, the success constitutes a state of reinforcement; that is, the goal of action unit completion has been achieved. The orientation value that is to be increased after successful action unit performance is given, for the five cases described, in Table 4.2.

After successful action unit completion, an increase takes place for each member in the group in the orientation which characterizes the group—that is, the orientation which has the largest value. The exception to this is case C in which the group's orientation is different from the

Table 4.2   Orientation Reinforcement

Case	Similarity of Maximum Orientation Value		Reinforce the Orientation which Characterizes
A	group = crew	= action unit	group
B	group = action unit	≠ crew	group
C	crew = action unit	≠ group	crew
D	group ≠ crew	≠ action unit	group
E	group = crew	≠ action unit	group

crew's orientation. In case C the orientation that characterizes the crew is increased for each member of the group.

COMMUNICATION NETWORK

It is assumed that communications between group members are sufficiently critical to system efficiency as to require simulation. The model assumes that the system is best which (1) has the fewest stations that must communicate (station term); (2) has a communication link from each station to every other station (link term); (3) is balanced so that all stations have the same number of communication links going to them as coming from them (load uniformity term); and (4) has no stations that are unlinked with the system (isolated station term).

In addition the model recognizes that random noise in the system will affect communications efficiency and makes an appropriate adjustment to account for this effect.

ENVIRONMENTAL STRESS

The group simulation model conceives of psychological stress that is the result of two types of situations—confinement and emergencies. This differs considerably from the one- and two-man model in which the stress is primarily time induced. Torrence (1961) reported a theoretical curve that relates performance proficiency and stress. The curve shows performance proficiency to be relatively unaffected by slight increases in stress up to a given point; above this point an increase in stress produces an increase in performance proficiency until a peak is reached; further increases in stress decrease performance proficiency.

The second aspect of stress, that caused by prolonged confinement, is based on data provided by Adams and Chiles (1961) and by Alluisi, Hall, and Chiles (1962). The work of these investigators indicated that for a 15-day confinement period performance proficiency increased during the first quarter of confinement and then slowly decreased. Following this decrease and when more than half of the total period of confinement had passed, efficiency again increased to almost a maximum and then leveled until the confinement was ended. These functions are incorporated in the model as quantified in Chapter V.

LEADERSHIP, ROLES, AND STATUS

In the model, leadership, roles, and status are considered to be extrinsically determined—for example, determined by the organizational rules and doctrine. Thus these variables are not considered as such in the current model. However, the aspect of supervisors' level of expectation for successful performance is simulated.

GROUP AND CREW SIZE

Total work crew size for the purposes of the present work is considered to be an independent variable. One of the major purposes of the model is to allow for independent variation of the crew size and composition in order to evaluate the effects of these changes on performance effectiveness. Group size in the model is dependent on the needs of the action unit and in the normal situation a complete complement is assigned by the computer. If for any reason (other assignments, sickness, etc.) an insufficient number of properly qualified crew members is available for assignment to the group, the action unit execution time is affected adversely.

TASK PERFORMANCE AND PERSONALITY

Task (action unit) performance is considered to constitute the dependent variable since the purpose of the model is to evaluate the effects on performance effectiveness of varying the parameters described. Although personality has not been included as a variable per se in the current conception, it is reflected by way of the values given for individual orientation. It might be noted that no assumption restricting personality to these variables is suggested but rather that the variables are considered to be of importance in forming personality traits.

SOCIAL-CULTURAL CHARACTERISTICS

Like the leadership structure and the job role that one plays, cultural variations are not believed by the current authors to be sufficiently variant to be influential in a defined and structured work situation. Accordingly they are not directly considered in the model.

V

THE GROUP SIMULATION MODEL—QUANTITATIVE CONSIDERATIONS

BRIEF OVERVIEW

Certain variables pertinent to small-group effectiveness have been identified. In this chapter along with certain other variables they are to be mathematically related and woven into a logic that forms the backbone of the digital simulation model. A brief summary of the model precedes the more detailed description of the pertinent features.

As in the case of a dual-operator man-machine simulation the use of this model is also based on the high-speed digital computer that executes the calculations in accordance with the model's logic. The computer operates on source data concerning performance limitations, as well as personnel, mission, and equipment-related data. Prior to the use of the model, analyses are performed of the man-machine system under consideration. These analyses involve the equipment in the system to be analyzed, the mission to be simulated, and the characteristics of the staff which will man the system. Each major task, called an "action unit," that is to be performed is identified and certain specific required source data are compiled. These data are set in punched card or magnetic tape form, or both, for introduction into the computer, for which a computer program or instruction sequence has been prepared. As directed by its program the computer will sequentially simulate, according to the logic of the model, the performance of the crew's operations—performance of each action unit being simulated in turn. Figure 5.1, a macroflowchart of the model, shows the model's general segments and sequencing.

The first major computer program segment is data read in. As shown in Figure 5.1 these data are entered in five distinct sets. Next, the minimum crew size is determined by the computer on the basis of the mission (work load) data.

Before simulation of the crew's performance on each given action unit assignment, the computer must make certain determinations about the crew itself. The composition of the crew by various types of personnel (both primary and alternate technical specialties) and by level is determined according to the mission requirements and personnel criteria. In addition each crew member is given an identity by assigning within the computer memory, initial values to his orientation and his technical proficiency (skill measure).

The simulation of crew operations is performed for each day of the mission by performing arithmetic operations on action unit data taken either from the predetermined mission data or generated as a result of repairs because of simulated equipment failures. (Equipment repair action units are automatically generated on a random basis in proportion to the specified failure rates of major systems.)

The next segment of the model deals with selection of the working group (circle 5 of Figure 5.1). For each action unit the model requires the selection of the group to perform the action unit. Men are assigned to the group on the basis of the amount of time worked during the day (the least amount is preferred), their pay level (the lowest is chosen first), and their proficiency level (the highest is chosen first). Three alternate modes of selecting group members are provided as represented by the following cases:

1. Normal action unit—preference in selection is given to personnel to work in their primary specialty.
2. Training action unit—preference in selection is given to personnel in one of their alternate specialties.
3. Difficult action unit—only primary specialty personnel are selected.

The simulation of the action unit itself is the next major segment of the model (circle 30, Figure 5.1). This simulation aspect involves three phases as follows:

1. The execution time calculation determines the time that the group uses to accomplish the action unit. The calculation is based on the proficiency of the group members, the overtime load on the group members, morale, the number of men who are required for performing the action unit but who are unavailable and average time data given as input.
2. The group performance efficiency computation is a function of sepa-

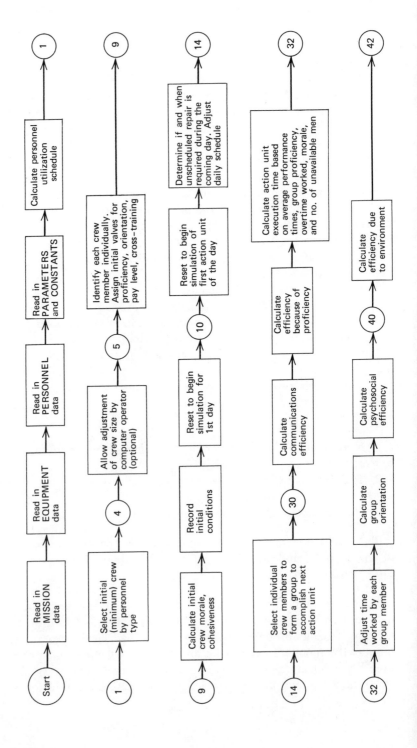

Start → Read in MISSION data → Read in EQUIPMENT data → Read in PERSONNEL data → Read in PARAMETERS and CONSTANTS → Calculate personnel utilization schedule → (1)

(1) → Select initial (minimum) crew by personnel type → (4) → Allow adjustment of crew size by computer operator (optional) → (5) → Identify each crew member individually. Assign initial valves for proficiency, orientation, pay level, cross-training → (9)

(9) → Calculate initial crew morale, cohesiveness → Record initial conditions → (10) → Reset to begin simulation for 1st day → Reset to begin simulation of first action unit of the day → Determine if and when unscheduled repair is required during the coming day. Adjust daily schedule → (14)

(14) → Select individual crew members to form a group to accomplish next action unit → (30) → Calculate communications efficiency → Calculate efficiency because of proficiency → Calculate action unit execution time based on average performance times, group proficiency, overtime worked, morale, and no. of unavailable men → (32)

(32) → Adjust time worked by each group member → Calculate group orientation → (40) → Calculate psychosocial efficiency → Calculate efficiency due to environment → (42)

66

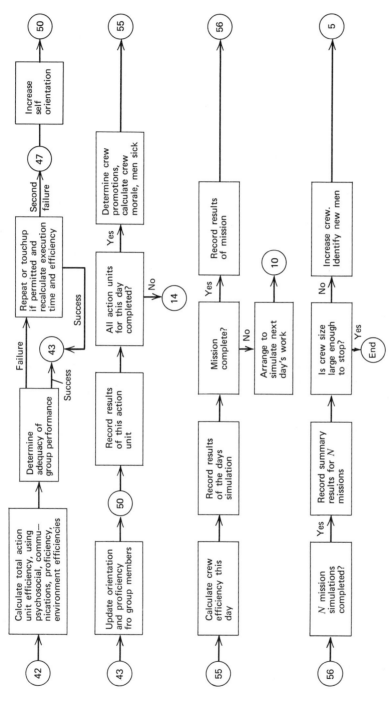

Figure 5.1 Macro flow chart of simulation. Note: Circled numbers correspond to circled numbers in Appendix B.

67

rate calculations for efficiency due to the following four elements:

(a) Communications—a function of the number of manned stations, link term, load uniformity term, isolated station term, and noise

(b) Group proficiency—a function of the current proficiencies of the group members in the specialties in which they are performing

(c) Environment—a function of an infrequent emergency and confinement stress

(d) Psychosocial interactions—a value based on a random process from a normal distribution whose mean is a complex function of the work group's orientation, the orientation of the entire crew, and that of the action unit

3. The adequacy of simulated performance is determined on the basis of the group performance efficiency and the efficiency of performance that could reasonably be expected from the group as estimated from the combined proficiencies of the group members. If performance is inadequate, the action unit may have to be repeated or performance may have to be "touched up" in order to meet minimum standards.

Certain action units may be postponed, or ignored completely, depending on circumstances. A nonessential action unit is postponed until the end of the next day if morale is below the morale threshold parameter or if its completion would require overtime work. Essential action units for which no suitable men are available are postponed until the beginning of the next day. Action units for which only a partial group can be assigned are initiated without delay but with a penalty in action unit execution time. In general the determination of priorities for selection of members for the group on the basis of avoiding overtime, work in a secondary specialty, postponing or ignoring an action unit based on input data and current conditions is implemented by relatively complex logic in the model.

The computer also maintains a record of the time of day (real time) at which each crew member worked last and the time of day at which each action unit is completed. These data are used to determine the actual time of day at which each action unit is assigned. The assignment is constrained by the requirements that (1) all selected group members must be available at the start of the action unit, (2) any designated prerequisite action unit specified by input data has been accomplished, and (3) the time-limited action units are completed by or terminated at the time specified.

Following action unit simulation the values for proficiency, orientation, and time worked are adjusted for each group member in accordance with the results of the performance. If desired, the results of each action unit simulation and the current values of pertinent variables are recorded for later review. The process is repeated similarly with new data and conditions for each action unit of the day sequentially. On a once-per-day basis a crew morale value is calculated from individual mission orientation values, individuals are selected for promotions in accordance with "policy," some are placed in or taken from a "sick" category to represent cases reflecting medical history data, cohesiveness indices are calculated, and a filtered end-of-the-day crew efficiency is calculated on the basis of efficiency of completion of each of the day's action units and the efficiency of the crew on preceding days. Summarized performance data and terminal conditions are recorded at the conclusion of the crew's work assignments for each day.

In this way the entire mission is simulated, day by day, with summarized mission results and end-of-mission conditions recorded for review. The entire mission is simulated in this fashion N times to average effects produced by certain random processes in the calculations and further summarized results are calculated and recorded.

Personnel are then automatically added to the crew by personnel type based on the overtime history of each type of personnel. This entire mission simulation process is further repeated after a determination of the next crew size to be simulated and records of results are made. A computer run is terminated when the crew, initially selected of minimum size, has been repeatedly incremented until it has reached a size sufficiently large to eliminate the need for overtime work by the crew or until the crew size has reached a predetermined limit.

The results recorded during the runs on magnetic tape are printed for visual review. The results yield such items of infomation as:

- System effieciency as a function of crew size and mission duration
- Morale and cohesiveness during the course of the mission
- Time spent in repairs
- Frequency and duration of overtime
- Manpower shortages by personnel types as a function of crew size
- Proficiency of each staff member during and at the end of the mission
- Variations in orientations during and at the end of mission as a function of crew size
- Frequency of necessity to skip or ignore nonessential action units
- Frequency of necessity to perform work in an alternate specialty
- Average and daily manhour loadings by personnel type

Repetition of the mission with different parameter values such as morale thresholds, overtime thresholds, and a variety of other initial conditions is possible. The use of different equipments, longer or shorter missions, and different numbers of personnel by type may be planned and simulated. Each run yields its own records and printouts which the system analyst may use in order to compare alternate systems, crew compositions and sizes, and missions in an attempt to optimize predicted performance in view of doctrinal, financial, and technical constraints.

SYMBOLS AND LIMITS

A reference list of symbols used in the model is given in Appendix C together with a range of numerical values encountered. (An xxx, for example, indicates a three-digit range of values, that is, 000 to 999.) Limits of the principal subscript values are also given in Appendix C; they define the limits of the model. The model, as implemented by a computer program, has been prepared to accept data for a mission of up to one year, having 200 action units per day in a system of no more than 20 stations and manned by crews of up to 150 men of up to 30 personnel types who operate up to 35 equipment systems.

In order to provide a useful tool and in view of the changes that may be expected in systems concepts in the future the goal was set to provide a flexible logic for the model but not based on too detailed a representation. Thus action units are selected to represent major work unit in tenths of hours. Only major equipment systems and crew operating stations are simulated. In this way the gross aspects of the social man-machine environmental complex are portrayed in an integrated, summary situation utilizing the model to yield results over a wide range of variables with relatively little effort in inserting new data into the simulation.

In order to gain a further perspective of the model and its concepts it is useful to review those features *not* included in the present approach. The following list indicates such features intentionally excluded:

1. Geographical location of the system during the mission
2. Destruction of the system or damage to the system other than normal equipment malfunctions that can be repaired by the permanent staff
3. Physical size and weight of system
4. Death or lay off of staff members during a mission

DETAILED MODEL LOGIC

A detailed discussion of the model and its features will now be presented. During the presentation reference will be made to the detailed functional flowchart of the model included in Appendix B. This level of detail provides a variety of important illustrations of techniques applicable in general model development and documentation.

Five data sets are used by the model: mission data, equipment data, personnel data, parameters, and constants. Read in of these data sets is called for and detailed in the first portion of the flowchart in Appendix B. Separate data sets are called for in order to facilitate a variety of simulation runs. Thus, for example, several different missions can be simulated sequentially using the same equipment by changing only the mission data. Manning and skill levels may be varied by reruns in which only the personnel data are changed without modifying mission or equipment characteristics. This feature is, of course, independent of the ability of the analyst to select any one or more specific input parameters for investigation by repetitive simulations on the computer. Since the object of the model is to evaluate many systems, different runs can be processed by the model for each of the five types of data sets making necessary the identification of each data set used. An identifier or description—usually the name of the mission, equipment key names, or the like—is assigned to every data set and accompanies the data to which they refer.

MISSION DATA

The duties of the crew on each day d of the mission are analyzed sufficiently to be delineated in list form. For each action unit u in each day's assignments of the crew the following data are required*:

1. The average time $\bar{h}_{ud}$ in tenths of hours, required to complete action unit u on day d

2. An indication of whether or not h_{ud} represents a fixed time period, a variable time period, or a time-limited period. A time-limited period is one that must end by a specified time of day regardless of the starting time (variable = 0, fixed = 1, limited = 2)

3. An indication of whether or not the action unit performance is to be repeated or improved in the event that performance efficiency does not come up to expectation (no repetition = 0, repetition or touch up = 1)

4. An indication of whether or not communication between stations is required on this action unit (yes = 1; no = 0)

5. The carry-over code that specifies the importance of the action

* A sample set of mission data for a hypothetical mission simulated is given in Table 6.5.

unit:

$$essential = 1$$
$$postpone\ to\ avoid\ overtime = 2$$
$$postpone\ if\ crew\ morale\ is\ below\ the\ threshold = 3$$
$$ignore\ to\ avoid\ overtime = 4$$

6. The orientation values of the action unit:

benefit to individual, s_{ud}
benefit to crew, c_{ud}
benefit to mission, m_{ud}

7. The type of action unit—an indicator of the preference in selecting group members:

normal, prefer primary specialty = 1
training, prefer alternate specialty = 2
difficult, insist on primary specialty = 3

8. A prior action unit number q_{ud} that must be completed before action unit u can be initiated.

9. Indicators e_{ud} to specify for each equipment system whether or not it is required to accomplish the action unit (yes = 1; no = 0 for each equipment, e).

EQUIPMENT DATA

The model assumes the compartmentalization of the total simulated system into x stations. The various equipments or equipment systems that enable the crew to accomplish their assigned action units are located within each station. The relative placement and sizes of the stations are not simulated by the model, although conceptually it is desirable for the system planners to have a station or equipment placement plan in mind for each simulation run.

The station communication matrix provided to the computer for the simulation describes for each station whether or not there is communication from that station to each other station. Facility for either direct verbal communication or communication over radio or telephone and possibly visual communication would be considered to represent the availability by interstation communication. Intrastation communications are assumed for each station by visual if not audible means. For each matrix element the numeral one is used to indicate the availability of communications and a zero is used to indicate the lack of communications (diagonal entries are zero arbitrarily).

The balance of the equipment data set is given in sequence by equipment number. For each equipment the following information is required by the model:

1. The equipment failure rate per day, the probability of failure of equipment e in a 24-hour period
2. Average repair time in tenths of hours
3. Station number at which the equipment is located
4. The number of personnel required to operate the equipment. This is given for each personnel type
5. The number of each type of personnel required to repair the equipment

The equipment related data regarding failure rate and repair time are usually available from equipment reliability and maintainability predictions or from records of experience with similar systems. If these sources are employed (and there may be no other sources for the data), a degree of dissimilarity may exist between the input data and the operational system that the model is attempting to predict. Thus nonconservative estimates of failure rates and average repair times are best selected.

PERSONNEL DATA

The third data set prepared for the computer simulation contains information on the personnel who will compose the crew of the confined system. Personnel are categorized by type t. Any one personnel type includes all those familiar with the same general discipline— trainees to highly experienced persons. Some type numbers (from 1 to T_e) are reserved for line personnel; others (from T_e to $T_e + T_o$) are reserved for staff. These assignments are flexible with the limitation of a maximum of 30 types of personnel.

The model allows for each individual i of the crew to have a primary specialty associated with or equivalent to his primary type number. In addition since each crew member may be cross-trained to do the work of different personnel types, the model recognizes the possibility that each crew member may be cross-trained for one or two specialties other than the one designated by his primary personnel type. Thus the remaining personnel data required are:

- The probability of the average crew member having one, and only one, alternate specialty, $p_{a=1}$
- The probability of the average crew member having exactly two alternate specialties, $p_{a=2}$
- Number of personnel types allotted to line personnel, T_e
- Number of personnel types allotted to staff (supervisory), T_o

The personnel types with which an individual may be cross-trained are called his "first alternate" and "second alternate" specialties. In order to determine the alternate specialties of each crew member with simulated reality during the calculations, a table of cross-training probabilities is required. Each entry in the matrix whose maximum order is 30 is a probability of an individual of one type having been cross-trained to perform the functions of some other personnel type or specialty. Many entries are expected to be zero to reflect the fact that personnel in one specialty do not possess the training for performance in nonclosely associated specialties. Such data are usually available in personnel records or can be estimated from previous experience.

PARAMETERS

The following six parameters are used in the model:

- Crew size increment Δ—the number of men to be added to the crew between each set of N simulations
- Morale threshold M—a value between zero and one. When crew morale is less than M, certain designated action units (carry-over code 3) will be postponed and the time of performance of others will be lengthened. Values for this parameter in the range of 0.3 to 0.8 will generally yield logical results. The choice of the precise value to be employed is judgmental and depends on the analyst's conception of the system and personnel subsystem under consideration.
- The nominal number of working hours per day W—given in tenths of hours. If more time is required to complete the work of a given day than W, then the balance is considered to be overtime.
- The probability of an emergency situation occurring during an average day P. Emergency situations create a stress that is assumed to last for the duration of a single action unit.
- An initial pseudo-random number R_0—from which subsequent pseudo-random numbers are to be calculated as required.
- The number of iterations N to be completed for each crew size.

CONSTANTS

A variety of problem-dependent constants are required as input data to the model. Their values will not vary during a single simulation run. However, because these constants represent personnel, medical, and psychosocial information, it may be found desirable to alter their values as new and better information becomes available. For this reason the constants are made to be as readily alterable as the parameters within the model. Each constant is symbolized by a K with a superscript giving the sequence within the series. The values shown for some of the constants are representative of those employed in the tests of the model.

The first series of constants gives data on the proficiency for average crew members at the initiation of the mission. Their symbology is shown below:

	Primary Specialty	Both Alternate Specialties
Average proficiency	$K_1^1 = 0.8$	$K_5^1 = 0.6$
Standard deviation of proficiency	$K_2^1 = 0.1$	$K_6^1 = 0.1$
Minimum proficiency	$K_3^1 = 0.6$	$K_7^1 = 0.2$
Maximum proficiency	$K_4^1 = 1.0$	$K_8^1 = 1.0$

These constants represent the parameters of a normal distribution that is used later in the determination of the proficiency with which each crew member will begin the mission. Values for this series can generally be derived from personnel records. The cumulative probabilities of the occurrence of the pay grade level and technical specialties among the crew members are given in the second series of constants. These are used in the determination of the technical specialties and pay grade levels with which each crew member starts the mission. The third series of constants are the three weighting factors for the initial orientation of crew members. These constants represent averages around which individual crew members' orientations are later determined on a Monte Carlo basis at the beginning of each mission simulation. The values that must sum to unity are:

Self orientation $K_1^3 = 0.2$
Crew orientation $K_2^3 = 0.4$
Mission orientation $K_3^3 = 0.4$

The value of self-orientation is generally selected to be lower than the others since it can be assumed that most crews would be more crew and mission oriented than self-oriented. Five constants constitute the fourth series. These are coefficients in the equation for the total efficiency of an action unit (to be described later) corresponding to the following items:

Total efficiency $K_1^4 = 1.0$
Psychological efficiency term $K_2^4 = 0.3$
Communication efficiency term $K_3^4 = 0.2$
Proficiency efficiency term $K_4^4 = 0.3$
Environmental efficiency term $K_5^4 = 0.2$

The three constants in the fifth series are employed in the determination of crew sickness and promotions at the end of each day. Values are obtained from personnel and medical records.

$K_1{}^5$ probability of an individual being promoted in any given day
$K_2{}^5$ probability of an individual entering becoming sick on any given day
$K_3{}^5$ the average number of days of each sickness

The next series of three constants consists of coefficients used in the numerical filter of the end-of-day efficiency equation to provide a smooth series of crew efficiency values.

The final constant $K_1{}^7$ specifies the percentage of expected efficiency that must be achieved in order to consider an action unit successfully completed. The value was selected to be 0.95, indicating that an action unit would be considered to be performed successfully if the efficiency of performance was at least 95 percent of expected proficiency.

FORMULATION OF A CREW

We shall consider now that all required data are prepared and available to the computer. As a first step and on the basis of these data the computer will automatically form a crew. But it is obvious that the action units specified for the mission can be accomplished with varying degrees of success by various size crews. Questions that immediately arise are which crew size, then, should be simulated and what shall be its composition by personnel type? These questions are answered on the basis of a calculation by the computer of a personnel utilization schedule and the results derived therefrom.

Yet the possibility exists that the system analyst will want to simulate with a crew size or composition other than the ones that may be generated by the computer. This would be desirable, for example, in order to simulate a greater crew size than the maximum or a lower one than the minimum, for a post-run fill in of a crew size skipped over, or to account for unusually heavy repairs that are not included in the computer's crew-size calculations. To provide for this the system analyst is given the option of specifying the crew composition by personnel type with complete flexibility. This decision is controlled by switch 2 in the flowchart shown in Appendix B.

The calculation of the personnel utilization schedule is based on a computation of the number of manhours required of each personnel type for each day of the mission to accomplish all action units at average $(\bar{h}_{ud})$ speeds, neglecting emergency repairs. At least one man of each personnel type is assigned to the crew regardless of how low his projected workload may be. The number of men required to complete the work in a single day in which the largest number of manhours is required, is selected to be the number of men in a minimum crew. This minimum crew size is the initial crew simulated; larger crews are simulated performing the same mission in later computer operations by successively and automatically adding personnel of the most needed types. Before beginning any simulations, the model sets an upper limit to crew size equal to the number of men obtained by summing the largest number of man days required to accomplish the specified action units by each personnel type regardless of the day on which this largest requirement occurred.

Having minimum and maximum crew sizes, the composition of the minimum crew by personnel type is then determined on a proportional basis so that simulation may begin.

IDENTIFICATION OF INDIVIDUAL CHARACTERISTICS

The next segment of the model is devoted to the assignment of initial characteristic values to each of the crew members. Each crew member is given an identity by calculating values for the following variables:

- Crosstraining specialty or specialties
- Self, crew, and mission orientation
- Proficiency in each specialty
- Staff or line level

The values thus calculated are stored by the computer for each crew member and serve as the initial conditions under which the crew starts the mission. During the course of the mission some of these values will be altered as a result of crew performance and promotions and the computer will adjust them in memory accordingly. Each crew member is assigned a number i for his identification in the crew. These numbers are sequentially assigned by personnel type t as previously determined.

CROSS-TRAINING SPECIALTIES

Cross-training specialties are determined first. The value of $p_{a=1}$ and $p_{a=2}$ from personnel input data determine whether the average crew member is cross-trained in one or two secondary specialties other than his primary personnel type. In order to determine how many alternate specialties are to be assigned to each specific crew member the computer generates a pseudo-random number R_y equiprobable in the interval 0-1 calculated as the next in an effectively infinite series started with the parameter R_0. The individual i is considered to possess no cross-training if R_y is greater than $p_{a-1} + p_{a-2}$ or, alternately stated, i will be cross-trained in one or two alternate specialties with probability $p_{a-1} + p_{c-2}$ when this process is performed over many individuals. The same pseudo-random number R_y is then compared against p_{a-1}, the probability of having exactly one alternate specialty. If R_y is less than p_{a-1}, one alternate specialty is assumed for each crew member; otherwise i has two alternate specialties. Hereafter each pseudo-random number from a rectangular distribution will be labeled R_y with the understanding that each new use implies another different number generated from the sequence.

The model now provides for the selection of those alternate specialty types that are the one or ones to be associated with each crew member. Assuming that a given crew member is to be cross-trained in one or two alternate specialties the question that must be answered is which personnel types are to be assigned? For any given simulation this determination is also based upon a Monte Carlo approach constrained so that the cross-training probabilities given in the table of personnel data are maintained in the long run. It should be noted that immediately following the read in of personnel data a new cross-training probability table is generated from the given table. Each element of the new table $\bar{p}_{tt}$ is the sum of all the elements p_{tt} that appear to the left of the corresponding position in the cross-training probability by row. For example, assume that for a selected i the elements of the original table are $(0, 0.8, 0.6, 0.3)$ for four personnel types; then the cumulative elements p_{tt} are $(0, 0.8, 1.4, 1.7)$. To determine the first alternate specialty to be assigned t_t^1 the model calculates $R_y \cdot \bar{p}_{tT} = R_y (1.7)$. Then the alternate specialty selected is the minimum t for which $\bar{p}_{tt} \geq R_y \cdot \bar{p}_{tT}$. In this example if $R_y = 0.5$, $R_y \cdot \bar{p}_{tT} = 0.85$ and $t = 3$ is selected as the alternate specialty for the i under consideration since $0.8 < 0.85 < 1.4$. This is repeated for the second alternate specialty t_i^2 if it was previously determined that i was to have two alternate specialties. The process continues in sequence for all personnel in the crew.

ORIENTATION

The self, crew, and mission orientation of each crew member at the start of the mission is calculated next. Each orientation value will fall in the range from zero to one and the sum of the three must be unity. Initial values are determined in a Monte Carlo fashion but are constrained in the long run to be related by the weights given by the constants K_1^3, K_2^3, and K_3^3, which also sum to unity. Specifically three pseudo-random numbers are calculated and individual orientation values are determined. The implementation normalizes the values to fall in the unit range, to sum to unity, and to be weighted in accordance with the constants in the long run (see circle 8 of Appendix B).

INITIAL PROFICIENCY VALUES

Up to three proficiency values are then calculated for each crew member. The first is the initial proficiency P_i^0 at the start of the mission representing rated or tested proficiency of individual i in his own specialty. The value for P_i^1 and P_i^2 are calculated to identify the initial proficiency of each individual in his first and second alternate specialty (if any), respectively. Initial proficiencies have an average of K_1^1 for the primary specialty and K_5^1 for both alternate specialties. The computer calculates proficiencies for each i under the assumption that the proficiency of the men in the crew in their primary specialty is normally distributed around K_1^1 with standard deviation K_2^1; and for alternate specialties distributed around K_5^1 with standard deviation K_6^1. The computation of a reasonable and realistic specific value for the actual initial proficiencies is made on the basis of a random selection from a normal distribution. Again the computer accomplishes this selection by a Monte Carlo technique. Using pseudo-random numbers three different random deviates R_d are determined and the values of initial proficiency are calculated as follows:

$$P_i^0 = K_1^1 + R_d K_2^1 \qquad K_3^1 \leq P_i^0 \leq K_4^1$$
$$P_i^1 = K_5^1 + R_d K_6^1 \qquad K_7^1 \leq P_i^1 \leq K_8^1$$
$$P_i^2 = K_5^1 + R_d K_6^1 \qquad K_7^1 \leq P_i^2 \leq K_8^1$$

Thus an assumption explicit in the calculations is that the proficiencies in the first and second alternate specialties may be different for any given individual but are, in fact, equal in the long run.

PAY LEVEL

The pay levels of the crew are also calculated and assigned by the computer prior to actual mission simulation. The first personnel types T_e are assumed to be line workers and are assigned a level R_i from 1 to 6. The model determines this R_i in such a way that each line worker will have a single specific rating for the beginning of the simulation and so that the relative probabilities of the levels as indicated by the second series of constants will be achieved in the long run. To do this a pseudo-random number R_y is calculated and collated with the six constants, K_1^2 to K_6^2, which represent the cumulative probabilities of occurrence of the levels. Then the level is determined to be the subscript of the smallest constant that exceeds R_y.

For staff a similar scheme is followed using the constants starting with K_{11}^2. Personnel types from T_{e+1} to $T_e + T_o = T$ are considered staff and each individual crew member with a personnel type in this range is given a level R_i from 11 to 14 in accordance with the probabilities and constants. Thus only one R_y is required in the determination of R_i for each crew member.

In summary cross-training, orientation, proficiency, and level are determined for each crew member, as described, to begin mission simulation. Because these crew identity values are calculated by a random process, these initial conditions would not be expected to be the same on any two simulations—yet following the given formulas as constraints they are believed to represent reasonable and realistic selections for individuals and to represent values with the desired average conditions.

CREW MORALE AND COHESIVENESS

Initial crew morale and crew cohesiveness are then calculated from initial orientations. Crew morale M_c and cohesiveness indices I_c^1 and I_c^2 are calculated at the beginning of the mission and again at the end of each day's work by the crew. First, the self-orientation s_c, crew orientation c_c, and mission orientation m_c of the crew are determined as the sums of corresponding values over all crew members. The sum of the squares of the three sums is used to normalize the components, as shown at circle M/C of Appendix B.

This preserves the relationships $s_c + c_c + m_c = 1$ and $0 \leq s_c$, c_c, $m_c \leq 1$. This method of normalization using the square of the sums of orientation values has the effect of giving a greater influence to the large orientational variables and of decreasing the influence of small orientation values. The crew moral is then defined to be:

$$M_c = m_c(1 - \sigma_m); \qquad 0 \leq M_c \leq 1$$

where

$$\sigma_m = \frac{\sqrt{\sum_i (\bar{m} - m_i)^2}}{C} \quad \text{and} \quad \bar{m} = \frac{\sum_i m_i}{C}$$

Thus crew morale equals crew mission orientation when in the ideal case there is no dispersion in m_c values (that is, when every crew member has the same mission orientation). In other cases crew morale is moderated by its standard deviation in such a way that the larger the dispersion in mission orientation of crew members, the greater will be the decrease in crew morale from the m_c value. When the crew is homogeneously oriented to mission objectives, morale as defined by the model is highest and when the crew's feelings about mission or objectives cover a wide range, morale is low.

A morale deviation factor is also calculated at this point in the sequence for later use in determining the speed of group performance. This factor is a measure of the deviation of crew morale from the morale threshold:

$$0.7 \leq \frac{M}{M_c} \leq 1.3$$

Both cohesiveness factors $I_c{}^1$ and $I_c{}^2$ are functions of the variance of orientational values. In the first index the variance is calculated using the crew orientation value that is determined to be the largest of the three. If, for example,

$$c_c > s_c > m_c$$

then

$$I_c{}^1 = e^{-10\sigma_c{}^2}$$

where

$$\sigma_c{}^2 = \frac{\Sigma_i(\bar{c}_c - c_i)^2}{C}$$

In this way crew cohesiveness assumes a perfect value of one when variance is zero and it decreases as the variance of individual crew orientation values increases. In determining the second cohesiveness index the variance is calculated for all three crew orientational values. The largest of these, termed $(\sigma^2)^*$, is used to calculate this index, $I_c{}^2 = e^{-10(\sigma^2)^*}$. Thus in $I_c{}^1$ the selection of the maximum is performed at the level of crew orientation values, whereas in $I_c{}^2$ the selection is made at the level of the variance of crew orientation values. At this point all personnel-related initial conditions have been determined and a recording of these data is made on magnetic tape by the computer for subsequent printout and analysis.

CURRENT DAY'S WORKLOAD

During simulation the computer will retain a variable, $d = 1, 2, \ldots,$ D, indicating the number of the day currently being simulated. This day number is reset to indicate the start of the mission. The model calls for a determination of the exact workload and sequence for each day just prior to simulating that day of the mission. Action unit (mission) data supplied as input describe prearranged crew assignments for the complete mission but the input information does not indicate crew requirements resulting from sporadic equipment failures. This exigency is calculated by the computer prior to the simulation of each day's work in turn. Thus the answers to the following questions must be determined at this point each day:

- Which equipments are to be assumed to have failed in day d?
- What should be the sequence in which the crew performs the repair with respect to the given action unit assignments?

Both questions are answered by techniques involving pseudo-random processes. The basic requirement in determining the answers is that the failures in equipment occur at times that are in general accord with given failure rates (equipment data) when taken over many missions. The calculation is accomplished as follows. A pseudo-random number R_y is calculated for each equipment and then compared with the corresponding failure rate per day f_e. Those equipments for which f_e exceeds R_y are assumed to have failed during the day. Thus a failure is indicated with a probability of f_e. In such cases the place of the repair action unit in the sequence of action units is determined by taking the product of the total number of action units in the day and a new R_y. This has the effect of calculating a place for the repair that is equally likely to occur at any place in the daily sequence.

The computer then assigns appropriate data for the repair action unit corresponding to the mission data provided initially for each normal action unit as delineated at circle 11 of Appendix B. These data are then arranged for later selection in their proper sequence for action unit simulation.

The next operation is to accomplish the reset or initialization of certain computer storage locations that will be employed later during the simulation of crew operations.

SELECTION OF WORK GROUP

The computer now has data available for the day's work to be simulated—both action units specified in advance and repair action units. Each day's simulation consists of simulating the performance of each action unit in its proper sequence. The processing of the data for an individual action unit will now be described in terms of the general action unit u with the understanding that the same logic is followed for each u in turn.

The processing begins at circle 14 of Appendix B by resetting certain model variables required in the processing of u, followed by the selection of the next action unit to be simulated either a preassigned or repair action unit. The first test is to determine whether or not u should be skipped (ignored) due to low crew morale. This would occur only for action units when the carry over code is 3 and when the crew morale is less than the morale threshold parameter M.

Selection of the group of men who are to accomplish action unit u is the next step of the simulation. Each equipment required in the performance of the action unit is designated by $e_{ud} = 1$ in the mission data. For every such equipment specified the computer extracts the number of personnel of each type required to operate (or repair) the equipment n_{te}^1 (or n_{te}^2). For each such equipment a vector of T elements describes these personnel requirements by type. The computer sums these values by personnel type to obtain T sums, called b_{tu}—the number of men (mneumonic bodies) of each type needed to perform the work. The sum is then formed to determine the total group size B_u,

$$ B_u = \sum_{t=1}^{T} b_{tu} $$

The processing that follows has as its purpose the selection of individual crew members to meet these needs. Selected are b_{1u} personnel of type 1, b_{2u} of type 2, etc., for all personnel types. Processing of the data to make the selections is done sequentially for each personnel type.

For action units of types 1 and 3 (normal and difficult) selection begins at circle 15 of Appendix B. For type-2 action units (training jobs in which preference is given to selection of personnel in their alternate specialties) selection begins at circle 21 and will be discussed later. In the case of type-1 or type-3 action units personnel are selected on the basis of three criteria, given in order of importance below:

- Number of hours already worked w_{id} on this day d. Select the b_{tu} men required in the order of those who have worked the fewest hours first. If there are more than a sufficient number of type t personnel with equal w_{id} values, then select on the basis of the next criteria
- Level R_i. Select up to the total of b_{tu} men required in order of those at the lowest level. If there are still more personnel than b_{tu}

available for selection from the crew with equal w_{id} and R_i values, then select on the basis of the next criteria
- Proficiency $P_i{}^0$. Select the b_{tu} men having the highest proficiency and assign them to the group

This selection process is accomplished in the computer by a sort on w_{id}, R_i, and $P_i{}^0$ data for the given t in such a way that those at the "top" of the sorted list are selected first (circle 23 of Appendix B). It will be noted that a single sort on $100w_{id} + R_i + (1 - P_i{}^0)$ values places the available personnel in order for selection of the b_{tu} men required. The effect of this procedure is to determine crew composition primarily from qualified, available personnel who have worked least (in order to share the work load); secondarily, those who are the lowest level (to insure that work is done by the most junior man qualified); and, tertiarily, those of highest proficiency (to induce high efficiency).

Before actually making the selection final as described, a test based on the average performance time $\bar{h}_{ud}$ is made to determine if the performance of this action unit would require the potential group members to work overtime; that is, the computer determines whether or not $w_{id} + \bar{h}_{ud} > W$. The w_{id} value which is minimum (top of the sort) is used in this test since if overtime is required by this test, then all other personnel selected could also be expected to require overtime in order to complete the action unit. If the inequality is not satisfied, no overtime is involved and the individual who is at the top of the sorted list is confirmed to be a member of the group g which will perform the action unit. If this completes the confirmation of b_{tu} personnel, the processing continues similarly with the next personnel type. Otherwise the next highest i in the sorted list is examined as before until b_{tu} personnel are confirmed for group assignment.

If the inequality is satisfied, however, and overtime could be expected, then the logic is as follows:

- For normal action units (type 1) a flag (the t-processing indicator) is set to indicate that there are no further type-t personnel available for assignment to the group without working overtime. (This is true since it has been determined that even the i with minimum hours worked, would have to work overtime to complete u.) Thus cross-trained personnel are selected in preference to use of overtime in the primary specialties. Following the processing of all personnel types in primary specialties another pass is made through available crew personnel to select personnel for the group from those cross-trained in type t as an alternate specialty.
- For difficult action units (type 3) with carry over code 4 the

entire action unit is skipped since type-3 action units are to be ignored in preference to promulgating overtime work.

• For difficult action units with carry over code 2 the entire action unit is postponed until the completion of all required work the following day in accordance with the purpose of carry over code 2.

• For difficult action units with carry over code 1 (essential) or 3 (postpone only on the basis of low morale) the model calls for the assignment of the men to the group even though they may be required to work overtime.

A test is then made to determine whether all b_{tu} personnel have been selected for the given t. If not, the process repeats by examining the next crew member in the sorted list for selection of additional men of type t in accordance with the logic described. When all b_{tu} men have been assigned to the group, processing continues for the next personnel type, $t + 1$, by recycling to circle 16 of Appendix B. When all T types of personnel have been processed, the entire group will have been selected if no t-processing indicators were set during the selection process—that is, if sufficient personnel of each t were found who would complete the work without overtime.

If t-processing indicators had been set for one or more personnel types, the processing continues with the search for personnel who have been cross-trained in the personnel types that are still not scheduled for a full day's work. The logic for this beginning at circle 20 of Appendix B provides again for examining each personnel type in turn. For each t for which a t-processing indicator has been set the computer performs a sort on $100w_{id} + R_i + (1 - P_i^{1-2})$ as before for those personnel not already selected who are cross-trained in personnel type t (in either primary or secondary alternate specialty). The i at the top of the sort is tentatively selected for group membership and a test is made for impending overtime requirements for the cross-trained personnel. If overtime is not required, the individual tentatively selected is confirmed and the process is repeated for type t until all b_{tu} personnel have been assigned and continues for all t until the entire group is selected. However, if overtime is required of cross-trained personnel as well as of personnel of primary specialty, then the model selects and confirms for the group whatever man has worked the least time regardless of primary or secondary skills in type t, level, or proficiency. This process repeats as above until as much of the group has been selected as is possible and is limited by personnel availability.

Should the situation arise in which no more personnel of primary type t or cross-trained in type t are available, and the required manning

b_{tu} has not ben fully assigned, then the computer stores for later use a value K_{tu} to represent the number of men who were not available for assignment. If no men were available for the group—that is, if the group is null—then the simulation of this action unit is deferred until the next day at which time it will be accomplished prior to any other required action units.

For training action units selection of group personnel is performed in a manner similar to that used for cross-trained personnel in order by t (circle 20 of Appendix B). When overtime is required of cross-trained personnel, then personnel of the primary specialty are selected. If both would require overtime to complete the action unit, then the individual at the top of the sort is selected. Skipping or postponing type-2 and type-4 action units, respectively, is accomplished as before if overtime of cross-trained (not primary specialty) personnel is required. In this way the computer performs the bookkeeping of group selection and group members are tagged for latter reference by a binary group select indicator.

COMMUNICATIONS EFFICIENCY

Having determined group composition by individual the model proceeds to calculations of group performance. The efficiency of interstation communication E_{cu} is the first of four exponential efficiency terms calculated as a measure of man-machine system performance for each action unit. This term can be considered a measure of degradation from an ideal communications situation for which a value of unity is assigned. If no communication is required during the action unit as specified by the mission data indicator, then the communications calculation is bypassed and since no degradation is assumed, E_{cu} is set equal to unity. If it is required, however, the communications system is assumed to consist of the assigned men of the group operating at their assigned stations who communicate (or are linked) with each other in one or both directions within the confines of the system.

A communications system can be represented by a graph as well as by a matrix in which stations are represented by points and the communication paths by links between points. If A, B, and C are points and if A can send messages to B and C, and if B can send messages to A only, then this situation may be represented by either the following figure or by a corresponding matrix in which the ones indicate a link in the direction from the point corresponding to the letter at the left to the point corresponding to the letter at the top.

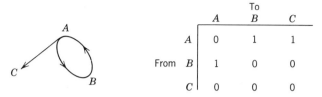

The raw data utilized in the calculation of the communications efficiency is the station communication matrix. Initially the computer forms a square submatrix from the communication matrix composed of only those rows and columns representing stations containing equipment that is manned during performance of the action unit. Thus if x_u is the number of such stations for action item u, the submatrix has dimensions x_u by x_u.

Four terms are then calculated in generating E_{cu}. The first term, called a "station term," is calculated to account for the degradation in communications efficiency anticipated as the number of stations increases. It assumes a maximum value of unity for the smallest system—that is, one consisting of two stations—and decreasing values for increasing numbers of stations. The term is $e^{-(1-2/x_u)}$, where x_u is the number of stations in u. For ten stations, for example, it has a value of 0.449.

The second or link term is to reflect better communications efficiency for a larger number of links in the system. A complete communications system is one in which there is a link in each direction between each station. Since for x_u stations there are $x_u(x_u - 1)$ possible links, the term is

$$\exp - \left[1 - \frac{1_u}{x_u(x_u - 1)} \right]$$

where 1_u is the total number of links—that is, the number of links in the submatrix. Numerical examples will be given later.

The third term aims to reflect load uniformity. The sum of the elements in a row of the submatrix represents the number of links coming from that station and the sum of the elements in a column represents the number of links going to that station. Ideally all row and column sums should be equal for uniformity of loading. Nonuniformity is a function of

$$\sum_{x_u} |s_r - \bar{S}| \quad \text{and} \quad \sum_{x_u} |s_c - \bar{S}|$$

where s_r is the sum of ones in row r $(r = 1, 2, \ldots, x_u)$, s_c is the sum of ones in column c $(c = 1, 2, \ldots, x_u)$, and $\bar{S}$ is the average of the column sums, $S = 1_u/x_u$. The uniformity term chosen then, having a maximum value of 1.0 is:

$$\exp - \frac{\left[\sum_{x_u} |s_r - \bar{S}| + \sum_{x_u} |s_c - \bar{S}|\right]}{21_u}$$

The final term, the isolated station term, measures the number of stations having no links coming to or going from them. This term is

$$\exp - \left[\frac{V_c - V_r}{2x_u}\right]$$

where V_c is the number of void (nonzero) columns and V_r is the number of void rows in the submatrix.

Several examples are given in Table 5.1 to illustrate the method of computation and to demonstrate the discriminative ability of the index. Graph parameters such as distances, associated numbers, existence of articulation points, etc., are obtained fairly easily by eye and yield intuitive measures of merit that have been quantized by the terms discussed.

Examples 1 and 4 are comparable except for a difference in size. Every point can communicate with every other point, although not directly in each case. The system represented in example 4 possesses reduced station and link terms. Examples 2 and 5 are also somewhat comparable. Here there is a reduction in value and examples 2 and 5 are rated lower than examples 1 and 4. The system in example 3 has higher station and link values than those of system 4. This is because of the difference in size and also because of the more complete connections existing in system 3. In system 4 the maximum path length between any pair is three links (average of two), while in system three the paths between A and C are at most two links long. The only improvement in system 4 is the uniformity of loading. System 4 is considered superior to system 5 because points B and D in system 5 cannot communicate with any other parts and A and C cannot receive messages. System 6 is superior to system 5 even though it has fewer links. The difference is in the link directions because A and B can both receive and transmit.

The model utilizes the four terms with equal weight and then considers a fifth element—that of noise in the communication system. Noise is assumed to be a random variable with a normal distribution. Thus the logic calls for the determination of a preliminary communications efficiency value that is the average of the four terms discussed. Then the precise value of E_{cu} for each action unit is determined by multiplying this average by a number selected from the normal distribution to represent noise. This number has a mean of 0.9, a standard deviation of

Table 5.1 Examples of Simple Communication Systems

Example	System Configuration	x_u	V_c	V_r	l_u	Station	Link	Terms — Uniformity	Isolation
1	see matrix below	3	0	0	3	0.72	0.61	1.00	1.00
2	see matrix below	3	1	1	3	0.72	0.61	0.52	0.72
3	see matrix below	3	0	0	4	0.72	0.72	0.72	1.00
4	see matrix below	4	0	0	4	0.61	0.52	1.00	1.00
5	see matrix below	4	2	2	4	0.61	0.52	0.37	0.61
6	see matrix below	4	1	1	3	0.61	0.47	0.69	0.78

Example 1 (triangle A→B→C→A):
```
    A B C  Σ
A   0 1 0  1
B   0 0 1  1
C   1 0 0  1
Σ   1 1 1  3
```

Example 2:
```
    A B C  Σ
A   0 1 1  2
B   0 0 1  1
C   0 0 0  0
Σ   0 1 2
```

Example 3:
```
    A B C  Σ
A   0 1 0  1
B   1 0 1  2
C   0 1 0  1
Σ   1 2 1
```

Example 4 (square A→B→C→D→A):
```
    A B C D  Σ
A   0 1 0 0  1
B   0 0 1 0  1
C   0 0 0 1  1
D   1 0 0 0  1
Σ   1 1 1 1
```

Example 5:
```
    A B C D  Σ
A   0 1 0 1  2
B   0 0 0 0  0
C   0 1 0 1  2
D   0 0 0 0  0
Σ   0 2 0 2
```

Example 6:
```
    A B C D  Σ
A   0 1 0 0  1
B   0 0 1 0  1
C   0 0 0 0  0
D   1 0 0 0  1
Σ   1 1 1 0
```

0.03, and is limited from above to a value of 1.0. This gives the effect of a degradation by noise of 0.9 on the average and a degradation between 0.81 and 0.99 (three sigma) 99.7 percent of the time. The total communications efficiency of the action units is:

$$
E_{cu} = \left\{ \frac{1}{4} \exp\left[-\left(1 - \frac{2}{x_u}\right)\right] + \exp\left[-\left(1 - \frac{1_u}{x_u(x_u - 1)}\right)\right] \right.
$$

$$
+ \exp\left[-\frac{\sum\limits_{x_u} |s_r - \bar{S}| + \sum\limits_{x_u} |S_c - \bar{S}|}{21_u}\right]
$$

$$
\left. + \exp\left[-\left(\frac{V_c + V_r}{2x_u}\right)\right] \right\} \{0.9 + 0.03R_d\}
$$

The negative exponential function was selected because it has the properties of assuming a unit value for a no-degradation case and assumes continuously declining values asymptotically to zero with increasing efficiency degradation.

EFFICIENCY BECAUSE OF PROFICIENCY

The second efficiency term, the efficiency resulting from the proficiency of the group members, is calculated as the average of the current proficiency values of the members of the group:

$$
E_{pu} = \frac{\Sigma_g P_i{}^a}{B_u}
$$

where $P_i{}^a$ is the proficiency of individual i in group g having the specialty in which he is selected to perform for action unit u. Individual proficiency values are determined at the beginning of the mission and are modified as a result of performance as discussed later. Thus this term will have the effect of increasing the crew performance efficiency when proficiency of group members increases and the reverse effect for decreasing proficiency.

ACTION UNIT EXECUTION TIME

The next calculation is the determination of the length of time required by the selected group of individuals to accomplish the action unit being simulated. If u is designated as a fixed-period action unit, the actual execution time h_{ud} is set equal to $\bar{h}_{ud}$ (mission data) and the processing continues with records adjustment at circle 32 of Appendix B. If the duration of performance is not specified in advance, the value of h_{ud} is determined as a function of the following five factors: (1) average performance time, $\bar{h}_{ud}$, (2) the deviation of the group proficiency from the nominal proficiency with which the crew starts the mission (execution time decreases as proficiency increases), (3) the amount of overtime that work group members have worked (work speed decreases as the work day lengthens), (4) the deviation of the crew morale from the morale threshold (work speed decreases as morale lowers), and (5) the number of required men who could not be assigned to the group due to unavailability of personnel of the desired types (execution time increases when fewer men than specified are available).

When average group proficiency (that is, efficiency because of proficiency) exceeds average initial proficiency, performance time decreases. This functional relationship is assumed to be linear and is limited so that the factor lies between 0.8 and 1.2 (see Figure 5.2):

$$\text{proficiency deviation factor} \qquad 0.8 \leq [1 - (E_{pu} - K_1{}^1)] \leq 1.2$$

The overtime degradation factor also assumes a linear relationship in that on the average for each hour that the group has worked overtime prior to starting work on an action unit, the factor will decrease by 0.02. For example, if a group of four men have worked a total of 16 hours beyond the standard work day of W hours, then the average overtime hours (4 hours) will result in a value for this factor of 0.92. The factor is given below and shown in Figure 5.2:

$$\text{overtime factor} \qquad 0.8 \leq \left[1 - 0.02 \left(\frac{\Sigma_g(w_i - W)}{B_u} \right) \right] \leq 1.0$$

The morale deviation factor previously discussed is calculated on a daily basis and used for each action unit. It is a measure of the deviation of the crew morale M_c from the morale threshold M. Morale higher than the threshold results in faster performance according to the relationship:

$$\text{morale factor} \qquad 0.7 \leq \left(\frac{M}{M_c} \right) \leq 1.3$$

The last factor is a measure of the effect of an incomplete group. If there are K_{tu} insufficient men of type t for assignment to the group on action unit u, then group performance time is as follows: performance time is unaffected if $\Sigma_t K_{tu} = 0$, it increases as a function of the number of unavailable men; when less than half the required group is missing, the factor is calculated according to the relationship:

$$\text{incomplete group factor} \qquad 1 \leq \left[2 \frac{\Sigma_t K_{tu}}{B_u} + 1 \right] \leq 2$$

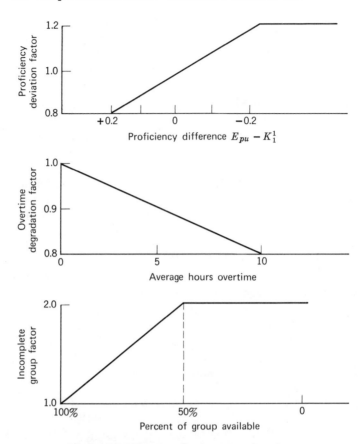

Figure 5.2   Factors affecting execution time.

When the condition arises that half or more of the group is unavailable, the factor is set equal to two; that is, performance takes twice as long.

Figure 5-2 displays the four relationships for execution time effects as conceptually developed by the present authors. The actual execution time h_{ud} of action item u by the group is calculated to be the product of the four factors (each suitably limited) and the average time $\bar{h}_{ud}$. The result is then itself limited as a function of the average execution time:

$$0.5\bar{h}_{ud} \leq h_{ud} \leq 3.0\bar{h}_{ud}$$

REAL TIME

The variable Z_i is used to denote the real time—that is, the time of day at which operator i completed his latest action unit assignment. A

Z_i value initially set to zero each day for every crew member is maintained by the computer. Since all group members who have been selected are to perform an action unit together, their respective Z_i values must be made to have equivalent values to reflect initiation of work in synchronism. The model actually calls for the logical equivalent of this; namely, the Z_i values of all group members are made equal after the action unit has been completed—that is, after h_{ud} has been determined. To accomplish this the computer determines the largest Z_i value of the group members Z^* that represents the earliest time of day when all group members were ready to work together.

The value of q_{ud} (mission data) is used to determine whether or not it is possible to begin u as early as time Z^*. Because q_{ud} indicates which action unit must be completed before starting u, the computer searches to determine the time Z_u at which action unit q_{ud} was completed. The action unit may then begin at the later of the two times Z^* (all men are available) or Z_{qu} (the prerequisite action unit is completed). The larger of these two is denoted Z^{**}. The group members are set in synchronism at the end of the simulated action unit performance by setting

$$Z_i = Z^{**} + h_{ud}$$

for each group member. A new Z_i value is stored to indicate the last real time each individual worked and a value of $Z_u = Z_i$ is stored to indicate the time when action unit u was completed.

The computer accomplishes the bookkeeping for the hours worked for each group member (circle 32 of Appendix B). Totals are kept for the primary specialty w_{id}^0, for alternate specialties w_{id}^{1-2}, and for total hours worked w_{id} by adding the action unit execution time to the three corresponding values for each group member. These totals are maintained for each crew member. If any new w_{id} value exceeds the maximum possible work day (22.0 hours) for one or more group members, they are tagged as "unavailable" for participation in the balance of the day's action units by setting their corresponding unavailability indicator for operator i, $a_{id} = 1$. A tally of all hours spent beyond 22 hours per day is kept and recorded later as unmanned station hours USH.

PSYCHOLOGICAL EFFICIENCY

Let us now consider the core of the simulation—how well the group performed as the result of psychosocial factors. Calculation of the psychosocial efficiency E_{pu} for action unit u is performed in accordance with the complex 16-term formulas shown at circle 35 of Appendix B. We note that an expected value of the psychosocial efficiency $\bar{E}_{su}$ is calculated first, followed by calculation of actual psychosocial efficiency E_{su} using a Monte Carlo approach.

The preliminary psychosocial efficiency calculation is based on three orientation vectors, each having three components. The vectors describe the group (g), the crew (c), and the action unit (u). The components of the orientation vector of each individual represent the extent to which that individual is self-oriented, crew oriented, or mission oriented (s_i, c_i, m_i) as previously discussed. The components of the group vector are derived using only those of the individuals in the group and represent the extent to which the group is self, crew, and mission oriented (s_g, c_g, m_g). Since a group is more than just the sum or the average of its parts, it is assumed that any difference in the averages of the vector components is accentuated by the group interaction. To calculate the components, therefore, the s_i, c_i, and m_i values over all group members are summed and these sums are raised to the second power for two-man groups or the third power for groups of three or more men. The modified components are then normalized; that is, each is divided by the sum so that $s_g + c_g + m_g = 1$.

The components of the crew vector (s_c, c_c, m_c) are derived in a similar way, except that the summation is over all the crew members including the group members and the power is always two.

The values of the action unit vector components s_u, c_u, and m_u (from mission data) are independent of each other in the range zero to one and represent the degree to which the action unit benefits the individuals of the group, the rest of the crew, and the mission, respectively.

We now recall the five cases of possible psychosocial facilitation considered in Chapter IV. It was desired to order these cases hierarchically from maximum support (case A) to minimum support (case E). The formula $\bar{E}_{su}$ was derived for use in the model to give values of 1.0, 0.8, 0.6, 0.4, and 0.2, respectively, for the five cases discussed. In the following formula for $\bar{E}_{su}$ primes indicate one's complements, for example, $s_c' = 1 - s_c$.

$$\begin{aligned}
\bar{E}_{su} = {} & 0.8(s_g s_c s_u + c_g c_c c_u + m_g m_c m_u) + 0.6(s_g s_u s_c' + c_g c_u c_c' + m_g m_u m_c') \\
& + 0.4[s_c s_u s_g'(c_g c_u)'(m_g m_u)' + c_c c_u c_g'(m_g m_u)'(s_g s_u)' \\
& + m_c m_u m_g'(s_g s_u)'(c_g c_u)'] + 0.2(c_g m_c s_u c_u' m_u' + m_g c_c s_u c_u' m_u' \\
& + m_g s_c c_u s_u' m_u' + s_g c_c m_u s_u' c_u' + s_g m_c c_u s_u' m_u' + c_g s_c m_u s_u' c_u') + 0.2
\end{aligned}$$

The formula was designed for the extreme cases (orientation elements of 0 and 1). To illustrate, $\bar{E}_{su}$ equals unity if, and only if, corresponding components of all three vectors are unity. In this case one of the three products of the first term is unity and the other two are zero, yielding a value for this term of 0.8. All products in the other terms are also zero, so $\bar{E}_{su} = 0.8 + 0.2 = 1.0$ as desired. The second term corresponds to case B, $\bar{E}_{su} = 0.8$. One and only one of its factors will be unity when there is a $(1, 0, 1)$ combination of components. All other variable terms will be zero, so $\bar{E}_{su} = 0.6 + 0.2 = 0.8$. The third term (with coefficient 0.4) corresponds to case C. The fourth term is associated with case D, where there is no coincidence, and the fifth term 0.2 encompasses case E, for which all variable terms are zero.

When the action unit vector has more than one component of unity, the effect of each would be additive without the introduction of a complement containing components in u. These occur in the third and fourth terms. On the other hand the use of complements of group and crew components alone, as in the second and third terms, represents only an algebraic simplification of the formula in combining two terms into one. For example:

$$s_g s_u s_c' = s_g s_u (1 - s_c) = s_g s_u (c_c + m_c) = s_g s_u s_c + s_g s_u m_c$$

The first equality is by definition; the second rests on the fact that $s_c + c_c + m_c = 1$. Calculations verify that the formula gives desired results for action unit vectors with more than a single one-unity element. To verify the calculation all 27 combinations of three different group, crew, and task orientations were calculated. Values selected for s_g, c_g, and m_g were $(0.7, 0.2, 0.1)$, $(0.1, 0.7, 0.2)$, and $(0.2, 0.1, 0.7)$. The crews, assumed less extreme, had the orientations $(0.5, 0.2, 0.3)$, $(0.2, 0.5, 0.3)$, and $(0.3, 0.2, 0.5)$, and action unit orientations s_u, c_u, and m_u were $(0.8, 0.1, 0.2)$, $(0.2, 0.8, 0.1)$, and $(0.1, 0.2, 0.8)$. The results given in Table 5.2 for the 27 combinations seem to indicate that the calculated values of $\bar{E}_{su}$ are acceptable from the point of view of logical consistency.

Having calculated the expected psychosocial efficiency $\bar{E}_{su}$ the actual psychosocial efficiency E_{su} is then determined. It is selected from a normal distribution in which $\bar{E}_{su}$ is the mean and the three-sigma interval is equal to the 0.2 value between the five cases discussed. Thus E_{su} is calculated by a random process and limited between zero and one as follows:

$$0 \leq [E_{su} = \bar{E}_{su} + 0.06667 R_d] \leq 1.0$$

ENVIRONMENTAL EFFICIENCY

The last of the four efficiency terms that measures the effect of the environment on the performance efficiency of the group is calculated as the average of the two values $E_{eu}{}^1$ and $E_{eu}{}^2$. The first reflects the effect caused by the occurrence of an emergency during performance of an

Table 5.2   Examples of Psychosocial Efficiency Calculations

s_g	c_g	m_g	s_c	c_c	m_c	s_u	c_u	m_u	E_{su}
0.7	0.2	0.1	0.5	0.2	0.3	0.8	0.1	0.2	0.688
0.7	0.2	0.1	0.5	0.2	0.3	0.2	0.8	0.1	0.500
0.7	0.2	0.1	0.5	0.2	0.3	0.1	0.2	0.8	0.451
0.7	0.2	0.1	0.2	0.5	0.3	0.8	0.1	0.2	0.635
0.7	0.2	0.1	0.2	0.5	0.3	0.2	0.8	0.1	0.558
0.7	0.2	0.1	0.2	0.5	0.3	0.1	0.2	0.8	0.478
0.7	0.2	0.1	0.3	0.2	0.5	0.8	0.1	0.2	0.660
0.7	0.2	0.1	0.3	0.2	0.5	0.2	0.8	0.1	0.510
0.7	0.2	0.1	0.3	0.2	0.5	0.1	0.2	0.8	0.496
0.1	0.7	0.2	0.5	0.2	0.3	0.8	0.1	0.2	0.506
0.1	0.7	0.2	0.5	0.2	0.3	0.2	0.8	0.1	0.641
0.1	0.7	0.2	0.5	0.2	0.3	0.1	0.2	0.8	0.529
0.1	0.7	0.2	0.2	0.5	0.3	0.8	0.1	0.2	0.436
0.1	0.7	0.2	0.2	0.5	0.3	0.2	0.8	0.1	0.685
0.1	0.7	0.2	0.2	0.5	0.3	0.1	0.2	0.8	0.513
0.1	0.7	0.2	0.3	0.2	0.5	0.8	0.1	0.2	0.478
0.1	0.7	0.2	0.3	0.2	0.5	0.2	0.8	0.1	0.635
0.1	0.7	0.2	0.3	0.2	0.5	0.1	0.2	0.8	0.558
0.2	0.1	0.7	0.5	0.2	0.3	0.8	0.1	0.2	0.553
0.2	0.1	0.7	0.5	0.2	0.3	0.2	0.8	0.1	0.457
0.2	0.1	0.7	0.5	0.2	0.3	0.1	0.2	0.8	0.652
0.2	0.1	0.7	0.2	0.5	0.3	0.8	0.1	0.2	0.510
0.2	0.1	0.7	0.2	0.5	0.3	0.2	0.8	0.1	0.496
0.2	0.1	0.7	0.2	0.5	0.3	0.1	0.2	0.8	0.660
0.2	0.1	0.7	0.3	0.2	0.5	0.8	0.1	0.2	0.513
0.2	0.1	0.7	0.3	0.2	0.5	0.2	0.8	0.1	0.436
0.2	0.1	0.7	0.3	0.2	0.5	0.1	0.2	0.8	0.685
0.3	0.3	0.4	0.3	0.4	0.3	0.1	0.1	0.1	0.784
0.3	0.3	0.4	0.3	0.4	0.3	0.2	0.2	0.6	0.527

essential action unit. The second is calculated once per mission day to reflect the effect of confinement on crew efficiency. In most applications the confinement effect will not be logically appropriate, in which case the $E_{eu}{}^2$ calculation is by-passed and the environmental efficiency based on $E_{eu}{}^1$ alone.

The calculation of $E_{eu}{}^1$ begins at circle 40 of Appendix B with the determination of whether or not an emergency occurs during performance of action unit u. It is assumed that an emergency situation exists whenever $R_y < P_u$, where $P_u = P/A_d$, P_u is the probability of an emergency during an action unit, P is the probability of an emergency per day (a parameter), and A_d is the number of action units to be performed on day d. If an action unit not essential (that is, carry-over code not 1), or if $R_y > P_u$, then $E_{eu}{}^1$ is set to unity to indicate no group efficiency degradation because of an emergency.

In the event of a simulated emergency the value of $E_{eu}{}^1$ is calculated as a function of stress in accordance with a relationship postulated by Torrance (1961). The reported relationship is separately supported by research of Harris, Mackie, and Wilson (1956). The function used to describe performance under stress was presented by Torrance as a "theoretical curve of a typical group performance under conditions of increasing intensity of stress." Hare (1962) also presents a theoretical curve of a similar shape relating effectiveness of performance over increasing stress with time. For use in the current model the data were replotted, scaled, and fitted by a straight line and a third-degree polynomial (Figure 5.3). For a stress value of zero to 2.8 the efficiency is approximated by the constant $3.5/6 = 0.5833$ (solid horizontal line in Figure 5.3). For stress values from 2.8 to 10.0 the following serves as the approximation:

$$\frac{-0.0208S^3 + 0.1736S^2 + 0.4628S + 1.300}{6}$$

In the model when an emergency is assumed to exist on any given action unit, a value for stress between 0 and 10 is generated equal to $10R_y$. With this known value of S the computer calculates the Figure 5.3 efficiency function. To obtain $E_{eu}{}^1$ itself the value obtained from the

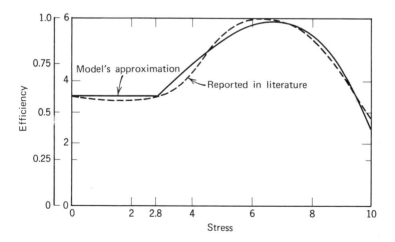

Figure 5.3   Efficiency as a function of emergency (stress).

model's approximation of Torrance's relationship is multiplied by the average of the two cohesiveness indices in order to encompass in a sense Torrance's conception of changes in affective linkage as a function of stress intensity. The change in affective linkage is suggested to follow the same function as that of Figure 5.3.

The second environmental efficiency factor E_{eu}^2 is that resulting from crew confinement and is calculated once per mission day. Figure 5.4, which represents the function relationship between performance efficiency and mission duration, was prepared on the basis of the data provided by Adams and Chiles (1961) and by Alluisi, Hall, and Chiles (1962). A series of straight lines fitted to the reported data are shown as a broken line in Figure 5.4. For purposes of the present model in which mission duration is flexible it was assumed that the reported function holds over any mission duration and that the same functional relationship holds when stated in terms of percentage of mission duration completed. For the computation, then, the percentage of mission completion is calculated as $Y = d/D$ to provide an argument for Figure 5.4. Then the factor E_{eu}^2 itself is determined according to the relation-

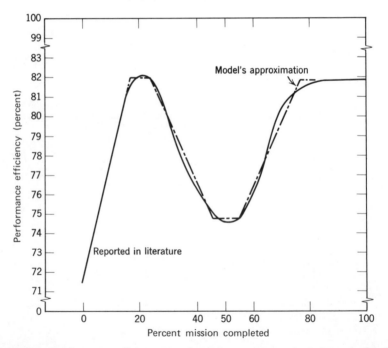

Figure 5.4   Performance efficiency as a function percentage of mission completed.

ship shown just before circle 14 in Appendix B. The combined efficiency resulting from the environment E_{eu}, both emergency and confinement, is calculated as the average of the two efficiency factors $E_{eu}{}^1$ and $E_{eu}{}^2$.

TOTAL EFFICIENCY

The total efficiency of the group in the performance of the action unit is calculated as a function of the four previously determined efficiencies, each multiplied by its appropriate scale factor constant:

$$E_u = K_1{}^4(K_2{}^4E_{su} + K_3{}^4E_{cu} + K_4{}^4E_{pu} + K_5{}^4E_{eu})$$

The scaling constants are incorporated to enable the manipulation of emphasis of relative importance of the different efficiencies with the restriction that they be selected so that the total efficiency E_u lies in the range zero to one. This total efficiency value is a major output of the simulation model and indicates just how well the group performed on the basis of the communication system at their disposal, their orientations, proficiencies, and environment.

PERFORMANCE ADEQUACY

The next question asked by the model is, "Did the performance of the group satisfy the requirements?" that is, was their efficiency E_u sufficiently high to call the action unit performance successful? The model's answer to this question is based on an assumption as to what could logically be expected of the men in the work group. It is assumed that the group, composed of men whose characteristics are known, could not be expected to do better than their combined proficiency levels permit. The level expected by their combined proficiencies has already been calculated to be E_{pu}. The person responsible for determining the adequacy of their performance could expect more or he might accept something less than this value before he would order a repetition or improvement in the work. The model expresses this "something more or less" by $K_1{}^7$ and calls the action unit successful if E_u is greater than or equal to $K_1{}^7 \cdot E_{pu}$.

In the event of success the computer proceeds to the next segment of the model, the adjustment of values of proficiency and orientation of group members based on this success (circle 43, Appendix B).

Should a failure be indicated, however, it is assumed that the same group members would be required to repeat or improve their work with the following two exceptions: If the fixed time indicator (mission data) indicates that the action unit execution time is a fixed value or if the repetition indicator (mission data) indicates that repetition or improvement is not appropriate for this action unit, then no opportunity is given to repeat or improve performance and processing continues at circle 50 of Appendix B ignoring adjustment of orientation and proficiency values. In the event of a failure on an action unit that is not so restricted, however, a single repeat attempt is made. The determination as to which is to be simulated, repetition or improvement, is made on the basis of another pseudo-random number. Repetition is called for only one time in ten, that is if $R_y \leq 0.1$; otherwise the same group members are ordered merely to "touch up" their performance with the aim of improving it.

In the case of a full repeat the computer recalculates the time of performance of the group h_{ud} as before. Since the group personnel and their characteristics have not changed (except for an increase in the hours worked), h_{ud} for the repetition will be the same as the original, except for a possible increase because of the overtime degradation factor. On the second try since the work group will have gained some familiarity with the nature of this action unit on the first attempt, it is assumed that communications efficiency has been improved to its maximum value of unity and that the psychosocial efficiency will be 0.02 (one case) higher than the value calculated on the initial attempt.

If the repetition of the action unit is successful on the recalculation of total efficiency as previously defined ($E_u \geq K_1{}^7 E_{pu}$), further processing continues as if the original attempt had succeeded. In the event of a failure on the second attempt group members' self-orientation values are increased as discussed later.

In the case in which improvement (touch up) in group performance is called for, the second execution time $h_{ud}{}^1$ is selected from a normal distribution within an average equal to one third of the actual unit performance time as previously calculated and a standard deviation of one ninth of h_{ud}. The recalculation of group efficiency in u for touch up is accomplished in the same way as described for action unit repeat.

MODIFICATION OF ORIENTATION AFTER SUCCESS

Social behavior is viewed in the present conception as both based on previous behavior and as a stimulus to future behavior. Thus at the completion of each action unit the orientation values of each work group member undergoes a change. Action unit success positively affects each work group member's orientations and proficiency which, in turn, will affect morale, cohesiveness, and efficiency on subsequent action units.

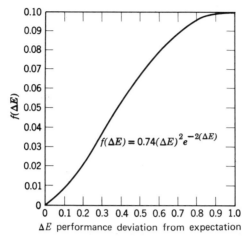

Figure 5.5 Functional relationship used in increasing the selected orientation variable.

The orientation values of each group member who participates in the successful completion of an action unit are redetermined after his work is performed. One of his three orientation values is selected as most salient to psychosocial efficiency as discussed in Chapter IV. For cases A, B, D, and E the increase will take place in the orientation that has the largest group orientation value s_g, c_g, or m_g. For case C the orientation value which is increased is the one having the largest crew orientation value, s_c, c_c, or m_c. The processing method for selecting the orientation to be increased is given at circle 43 of Appendix B.

The magnitude of the increase is determined by first calculating a value of ΔE, the group proficiency excess—the amount of action unit proficiency which was in excess of that required to satisfy the criterion of a successful action unit, $E_u - K_1{}^7E_{pu}$. The model then calculates $f(\Delta E) = 0.74(\Delta E)^2 e^{-2(\Delta E)}$, which relationship is shown in Figure 5.5. If, for example, s_i is the selected orientation, then the new value of s_i will be:

$$s_{i,u+1} = s_{i,u} + (1 - s_{i,u})f(\Delta E)$$

These formulas were adopted to maintain relationships such that increases in orientation will become smaller as the orientation values themselves become larger and increases in orientation will become larger the more the group efficiency on an action unit exceeds the acceptable orientation for success.

When the selected orientation is increased, the other two components are decreased by that same fractional amount which preserves the condition that the sum of the three equals a unit. If s_i is the orientation selected for increase, then the other two are adjusted as follows:

$$c_{i,u+1} = \left(\frac{1 - s_{i,u+1}}{1 - s_{i,u}}\right) c_{i,u}$$

$$m_{i,u+1} = \left(\frac{1 - s_{i,u+1}}{1 - s_{i,u}}\right) m_{i,u}$$

MODIFICATION OF ORIENTATION AFTER FAILURE

In the event of action unit failure on both attempts the self-orientation of each group member is increased. This is done to reflect the thinking that repeated failures will generate a retreat and a sense of self-justification on the part of group members. Because it is felt that the individual's mission orientation would be unaffected by this repeated failure, an equal decrease is made in crew orientation to offset the increase in self-orientation and no change is made in mission orientation. The formula for calculating the increase is the same as that used for increases in the event of success as discussed above, except that the absolute value of ΔE is used. Thus the amount of the increase in s_i is dependent on the amount by which the expectation exceeds performance.

INCREASE IN PROFICIENCY

The model conceives of an increasing proficiency situation. In the missions under consideration individual proficiency can increase but not decrease. Therefore after successful completion of an action unit the proficiency of each group member is increased in the specialty in which he actually worked during performance of u (circle 48 of Appendix B). The mission is assumed to be sufficiently short (weeks or months) so that no degradation of proficiency occurs in an unused skill. Nor is it assumed that failure of an action unit can degrade proficiency.

The proficiency updating scheme is based on the work of DeJong (1957) and of Crossman (1959). Of particular importance was the work of Blackburn (1936) and Crossman (1956), which indicated that performance proficiency on a task continued to increase even after 10,000 "trials."* Thus given average crew proficiency at the start of the mission

* Reviewers of an early manuscript of this text have indicated that even the seemingly conservative upper limit of 10,000 is modest and that for some tasks the 0.99 level might never be reached.

at 0.75, the formula was scaled so that a proficiency of 0.99 would be achieved after 10,000 action units in the same specialty. The proficiency of individual i in specialty a after action u is therefore given by:

$$P^a_{i,u+1} = P^a_{i,u} + \frac{(1 - P^a_{i,u})E_u}{2640}$$

SIMULATION OF SUBSEQUENT ACTION UNITS

This completes the processing for an action unit. Before proceeding in turn to subsequent action units, the computer will optionally record the results of the completed action unit. A sample of such output is shown in Table 5.3.

The time required to process an action unit is about 0.2 sec of IBM 7094 computer time when a full detail action unit printout is generated. In the event that only summary printouts are produced computer time per action unit was determined to be approximately 0.09 sec.

A DAY'S WORK IS DONE

The processing continues as described for the next action unit in turn and for all subsequent action units on day d. When the last action unit of a day has been completed, the computer makes the following determinations and calculations prior to recording the day's results: (a) crew promotions, (b) crew morale, (c) crew cohesiveness, (d) changes in sickness status and (e) crew efficiency.

Table 5.3   Sample Printouts of Action Unit Results

```
ACTION UNIT  12    DAY   1    ITERATION  1    REPAIR 0    TYPE 1    INDICATOR 0
GROUP PROF. 0.7585    ENVIR. EFF. 1.0000    CONFIN: EFF. 0.7855    MSH    0.
   GROUP ORIENTATION         CREW  ORIENTATION
  SELF  CREW  MISSION     SELF   CREW   MISSION
  0.076  0.692  0.232     0.193   0.395   0.413

FIRST ATTEMPT WAS UNACCEPTABLE    EXECUTION TIME   2.32    REAL TIME 14.32
COMM. EFF. 1.0000    PSYCHOSOC. EFF. 0.4080    TOTAL EFF. 0.6989

SECOND ATTEMPT WAS ACCEPTABLE    EXECUTION TIME   0.22    REAL TIME 14.55
COMM. EFF. 1.0000    PSYCHOSOC. EFF. 0.6460    TOTAL EFF. 0.7822
```

			GROUP STATISTICS							
I	TYPE	RANK	HOURS WORKED	PRIOR ORIENTATION			FINAL ORIENTATION			SPEC	FINAL PROF.
				SELF	CREW	MISSION	SELF	CREW	MISSION		
6	3	5	6.5	0.313	0.581	0.106	0.312	0.579	0.109	1	0.691
10	3	1	7.3	0.141	0.435	0.424	0.140	0.434	0.426	1	0.714
12	4	6	6.5	0.116	0.509	0.375	0.115	0.508	0.377	1	0.683
14	5	2	6.5	0.078	0.460	0.462	0.078	0.459	0.463	1	0.790
15	5	2	10.5	0.308	0.607	0.085	0.308	0.605	0.087	1	0.904
18	7	1	10.5	0.367	0.169	0.464	0.366	0.169	0.465	1	0.769

CREW PROMOTION

Crew promotions are based on the constant $K_1{}^5$, the probability of an individual being promoted on any given day. The value $CK_1{}^5$ is taken as an approximation of the probability of a promotion occurring in a crew of C men on any given day. The model calls for determining whether or not a man is to be promoted on day d by comparing a pseudo-random number R_y with $CK_1{}^5$. An arbitrary, but nonlimiting, assumption is made that no more than one individual may be promoted on a given day. A man is promoted if $R_y \leq CK_1{}^5$. If promotion is indicated by this test, the man with the greatest proficiency in his primary specialty is selected for promotion. Should the man selected already be in the highest staff or line-level categories, no promotion is made on that day. Following promotion of i an indicator is set to prohibit further promotion of the same individual during the balance of the mission. No adjustment of proficiency is considered necessary on a promotion since the man's operational responsibilities are assumed to remain related to action unit performance.

MORALE AND COHESIVENESS

Crew morale M_c and cohesiveness indices $I_c{}^1$ and $I_c{}^2$ are determined as previously described to represent end-of-day conditions.

SICKNESS

The model next makes a determination of whether or not any crew members must be considered unavailable because of sickness. All crew members are said to be available at the start of the mission, reflecting the assumption that only fit men will be assigned to begin the mission. Calculations are made to determine the following data for the next day: (a) the number of sick men, (b) the identity of sick crew members, (c) the duration of unavailability (days), and (d) orientation adjustment for sick personnel.

The number of men sick on a given day is assumed to be a Poisson distributed variable with a mean of $K_2{}^5 \cdot C \cdot M/M_c$, where $K_2{}^5$ is the probability of a man being sick on any given day, C is the number of men in the crew, M is the morale threshold, and M_c is the crew morale.

A pseudo-random number from a Poisson distribution with the mean as described is calculated and the integer selected represents the number of sick crew members. Thus the higher the crew morale, the fewer the number of sick personnel up to the point when the crew morale equals the morale threshold. In cases in which crew morale is less than the threshold M the model provides up to a fourfold increase in sickness from that which would be predicted at the threshold value.

To determine which individual is sick the model forms the product $R_y C$. The integer part of this product, which lies between zero and the maximum crew size, is used to specify the sick individual. This is repeated as many times as necessary to identify the number of men specified to be sick. The duration of the sickness a_{id} is determined sequentially and independently for each sick individual previously identified. A pseudo-random number is selected from a Poisson distribution with mean $K_3{}^5$ to represent the average duration of sickness in days a_{id}. Any man having a nonzero a_{id} value is considered unavailable and is not selected for work-group assignment in the next a_{id} days. The counting of a_{id} values is accomplished once each mission day by reducing each a_{id} value by one.

The model increases the self-orientation for each sick man thus reflecting the impact of the illness on crew morale and cohesiveness. This increase is calculated as described previously (circle 44 of Appendix B), where:

$$\Delta E = \frac{(1 - s_i)a_{id}}{10}$$

Thus the increase in self-orientation is dependent on the present self-orientation and the duration of the sickness (up to 10 days). The increase in s_i is specified for the sick individual at the expense of his m_i and c_i, which are reduced proportionately.

CREW EFFICIENCY

The average efficiency over the day's operations is determined as the final calculation of daily results (see circle 55 of Appendix B). The average efficiency E_d includes performance efficiency demonstrated only on essential action units. This restriction is made because a final calculation of efficiency on the important work of the day is desired. In order to gain some smoothness of results with respect to efficiency trends the computer performs a filtering operation on E_d using weighted values from prior days. A second-order difference equation accomplishes the action of the desired digital filter; the result is a smoothed value of efficiency $\bar{E}_d$ at the end of the day d equal to $K_1{}^6 E_d + K_2{}^6 \bar{E}_{d-1} + K_3{}^6 \bar{E}_{d-2}$. The values of the constants must be selected so as to yield adequate filter stability and response:

$$K_1{}^6 = 1 - K_2{}^6 - K_3{}^6$$
$$0 < K_2{}^6 < 2$$
$$-1 < K_3{}^6 < -\frac{K_2{}^6}{4}$$

SUMMARY RESULTS

Having completed all required processing for an entire day's operation the computer writes summary results on magnetic tape for later printing and analysis. A wide variety of data is recorded as shown following circle 55 in Appendix B. Some of the results represent summary tabulations and calculations of daily averages. These include, but are not limited to, average overtime, repair hours, number of successful and unsuccessful action units completed, and end-of-day conditions such as crew morale, efficiency, cohesiveness, and orientation.

At this point the entire process is repeated utilizing action unit data for the next day in succession by returning to circle 10 of the flowchart. Simulation of each day's crew operations continues in turn until on reaching the end of the last mission day the mission is considered to be completed.

WHEN THE MISSION IS OVER

At the conclusion of the mission, another set of summarized end-of-mission data is written on magnetic tape for subsequent analysis. At this level interesting and useful information emerges representing single-mission simulation. A sample iteration summary record is displayed in Table 5.4. Since many of the calculations in the model involve stochastic processes, no two repetitions of a single mission can be expected to agree exactly. To achieve reasonable and stable results it is necessary to repeat applications of this method several times. Thus the simulation of a mission is repeated N times with exactly the same conditions and data, except for the initial pseudo-random number. The count from one to the input parameter N is termed the iteration number.

Following N iterations of the entire mission, the final and most significant data are written on magnetic tape by the computer. These data are primarily averages of end-of-mission conditions averaged over the N missions. Among the results are:

1. Average crew efficiency at end of mission
2. Average crew morale at end of mission

3. Average crew orientation at end of mission
4. Average hours worked per day by type of personnel
5. Average crew cohesiveness at end of mission
6. Average hours spent in repair
7. Average proficiency of crew at end of mission

Table 5.4 Sample Results of One Complete Mission Iteration

```
RESULTS OF ITERATION NUMBER  1 OF ENTIRE MISSION   DURATION  10 DAYS   CREW SIZE  33
     CREW EFF. 0.8151   CREW MORALE 0.471   COHES. INDEX /1 =  0.74  /2 =  0.66
     AVG. HRS. WORKED  7.7   AVG. O/T HRS.  0.2   AVG. UNUSED HRS.  4.5   AVG. HRS./REPAIR  0.7
     MSH      9.2   SICKDAYS   18   WORKING HRS./DAY 12.0   MORALE THRESHOLD 0.400   MEN UNAVAILABLE   2
     AVG. HRS. WORKED/DAY - PRIME   7.2   ALTERNATE  0.5   TOTAL  7.7
     CREW ORIENTATIONS    SELF  0.128   CREW  0.301   MISSION  0.572
     PERSONNEL TYPE    AVG. HRS. WORKED    NO. OF MEN
          1                 5.1               3
          2                 3.8               1
          3                11.9               7
          4                 6.2               2
          5                 4.4               2
          6                 0.8               1
          7                 9.2               3
          8                 5.9               1
          9                11.3               2
         10                 4.8               2
         11                 4.5               1
         12                 5.0               2
         13                 4.8               2
         14                10.2               4
          INDIVIDUAL STATISTICS
     I   RANK   PROM        ORIENTATIONS            PROFICIENCIES
                          SELF   CREW  MISSION    PRIME   ALT1    ALT2
     1     4    0        0.150  0.387  0.463     0.658   0.512   0.
     2     1    0        0.142  0.333  0.524     0.588   0.346   0.
     3     1    0        0.386  0.231  0.383     0.886   0.324   0.364
     4     1    0        0.340  0.429  0.230     0.614   0.540   0.504
     5     1    0        0.247  0.185  0.568     0.845   0.418   0.
     6     5    0        0.246  0.483  0.271     0.694   0.515   0.
     7     1    0        0.318  0.044  0.639     0.744   0.344   0.416
     8     2    0        0.312  0.046  0.641     0.810   0.623   0.
     9     1    0        0.524  0.127  0.349     0.715   0.418   0.
    10     1    0        0.112  0.384  0.504     0.717   0.576   0.
    11     4    0        0.079  0.546  0.375     0.734   0.524   0.
    12     6    0        0.127  0.365  0.508     0.683   0.428   0.
    13     4    0        0.004  0.699  0.297     0.609   0.444   0.
    14     2    0        0.135  0.428  0.437     0.790   0.618   0.
    15     2    0        0.306  0.602  0.092     0.904   0.      0.
    16     1    0        0.109  0.337  0.554     0.627   0.495   0.
    17     1    0        0.276  0.154  0.571     0.744   0.469   0.
    18     1    0        0.232  0.101  0.667     0.772   1.496   0.251
    19     1    0        0.288  0.224  0.489     0.745   0.461   0.
    20     2    0        0.047  0.831  0.122     0.749   0.457   0.
    21     1    0        0.256  0.466  0.278     0.686   0.521   0.
    22     2    0        0.027  0.738  0.235     0.830   0.453   0.364
    23    13    0        0.212  0.300  0.488     0.661   0.444   0.
    24    11    0        0.342  0.251  0.407     0.810   0.367   0.
    25    11    0        0.255  0.636  0.110     0.580   0.      0.
    26    12    0        0.143  0.121  0.737     0.791   0.489   0.
    27    12    0        0.060  0.102  0.837     0.644   0.470   0.
    28    11    0        0.220  0.280  0.499     0.583   0.559   0.
    29    11    0        0.188  0.268  0.544     0.891   0.550   0.
    30    12    0        0.184  0.285  0.530     0.757   0.482   0.
    31    13    0        0.240  0.170  0.590     0.828   0.518   0.
    32    12    0        0.231  0.194  0.576     0.752   0.578   0.
    33    13    0        0.361  0.140  0.499     0.752   0.577   0.

AVERAGE PRIME PROFICIENCY  0.733      RANDOM NO. = 177743460242
```

OTHER CREWS

The balance of the model's processing is concerned with the selection and formation of larger crews to be simulated. The entire process of performing N simulations is repeated with an increase in Δ persons to the crew, where Δ is given as an input parameter. This would normally be continued until the crew reaches the maximum crew size, which was calculated in connection with the personnel utilization schedule prior to the first iteration.

Before adding to the crew, however, the model's logic calls for determining whether simulations should be terminated even prior to reaching that crew size. The average hours that were worked in the last N iterations are searched for each personnel type. If the largest of these is less than W (indicating that on the average no overtime was required by the longest-working personnel type), then the simulation run is terminated. Otherwise Δ is added to the crew and a determination is initiated (circle 57 of Appendix B) as to which personnel types are to be added. Men are added whose specialties represent the most needed personnel types. The supplementation is performed by successively adding one man of the personnel type that has worked most overtime per man in the last N iterations. This is repeated until Δ men have been added. The possibility of adding two men of the same type is allowed and is provided for by special processing of residual overtime averages. When the crew is up to its new complement, another N iterations with the revised crew are automatically initiated by returning to circle 5 of the flow chart.

The run of all crew sizes is thus terminated when the crew size reaches the maximum C^* or when no personnel type is required to work overtime—whichever comes first. The former case is identified by the computer printout of MAXIMUM CREW SIZE REACHED and, in the latter case, is indicated by MAXIMUM HOURS WORKED LESS THAN NORMAL.

By providing for increases in crew strength the effects of a variety of crew sizes on the same mission may be obtained and all mission summary data may be analyzed as a function of crew size.

VI

TESTS OF GROUP
SIMULATION MODEL

Chapter VI presents a hypothetical mission together with associated data selected to enable sensitivity testing of the group simulation model. These data were selected so as to result in a set of conditions that does not reflect any specific operational or planned system yet possesses properties similar to missions that may be of some practical interest. Although the exercise of this mission on the computer cannot be represented as a validation of the model against criterion data, the results are important for illustrating the model's internal coherency, the directions and rates of change (trends) for the model's variables, the operational features of the model, and its scaling, format, and general content.

Then the results of applying the model to an actual Navy system are described. In this application, which represents the results of a test of the predictive validity of the model, the ability of the model to predict required manning, morale, cohesiveness, efficiency, man hours required, and the like, for certain aspects of a nuclear, fleet ballistic missile submarine is investigated.

THE HYPOTHETICAL SYSTEM SIMULATED

For the sensitivity test of the model we conceive an underwater craft whose function is that of long-term isolated alert at alternating, stationary submerged points. The system's only function is retaliatory delivery of nuclear missiles on receipt of proper instructions.

Five stations in which men perform equipment operation and maintenance are defined within the craft. These stations include the engine room, the missile area, a command and control area, the medical area or infirmary, and the galley and crew recreation area. Attention should be drawn to the fact that no torpedo, self-defense equipment or nonoperating stations—for example, quarters—are simulated. The station communication matrix for this assumed configuration is shown in Table 6.1. In Table 6.1 a row-column intersection entry of ONE indicates that communication is possible between the stations indicated; a ZERO indicates no such communication is possible. Note that the command and control area is able to communicate with any other station.

Table 6.1   Station Communication Matrix for Hypothetical System

		TO Station Number					
		1	2	3	4	5	Station Name
FROM	1	0	1	1	0	0	Engine room
	2	1	0	1	0	0	Missile area
Station	3	1	1	0	1	1	Command-control area
Number	4	0	0	1	0	0	Sick bay
	5	0	0	1	0	0	Galley

Twenty-two major pieces of "equipment," which may require operation, or maintenance, or both, during the mission, are identified in Table 6.2. The list of the equipments also includes estimated failure rate, estimated average repair time, station location number, and equipment manning requirements for the operation and repair of each equipment. It can be seen from these data that the reactor, the reactor controls, the engine, and the generator are included in the engine room. Additionally, the two missile launchers are considered to be independent.

PERSONNEL

Fourteen personnel types are called for in order to operate and repair the equipment. In Table 6.3 the probability of cross-training between the nine types of enlisted men (line) and the five officer types (staff) is specified; the probability of having one alternate specialty was selected to be 0.6 and the probability of having two alternate specialties was 0.2.

PARAMETERS AND CONSTANTS

The parameters and constants are given in Table 6.4. Several runs were made during the sensitivity tests; each run contained a different set of constants and parameters. The first set, called set A, was a standard against which results of the other sets were compared. In set A 12 hours were designated as the normal workday and two men were added

Table 6.2 Equipment Data

Equipment No.	Failure Rate/Day	Average Repair Time	Operate 1	2	3	4	5	6	7	8	9	10	11	12	13	14	Repair 1	2	3	4	5	6	7	8	9	10	11	12	13	14	Equipment Nomenclature	Station No. and Name
1	0.14000	01.5	0	0	1	0	0	0	0	0	0	0	0	0	0	0	0	1	2	0	0	0	0	0	0	0	0	0	0	0	Reactor plant control	1 Engine room
2	0.00500	02.5	0	0	0	1	0	0	0	0	0	0	0	0	0	0	0	0	0	3	0	0	1	0	0	0	0	0	0	0	Steam engine	
3	0.00500	02.5	0	0	0	0	1	0	0	0	0	0	0	0	0	0	0	0	0	0	2	0	0	0	0	0	0	0	0	0	Generator	
4	0.03000	00.8	0	0	0	0	1	0	0	0	0	0	0	0	0	0	0	0	0	0	2	0	0	0	0	0	0	0	0	0	Alternate power supply	
5	0.00100	03.0	0	0	1	0	0	0	0	0	0	0	0	0	0	0	0	0	2	0	0	0	1	0	0	0	0	0	0	0	Reactor	
6	0.01000	03.0	0	0	0	0	0	0	1	0	0	0	0	0	0	0	0	0	0	0	2	0	0	1	0	0	0	0	0	0	Environment control equipment	
7	0.03000	02.0	0	0	0	0	0	0	0	0	0	0	0	0	0	0	0	1	0	0	0	1	0	0	0	0	0	0	0	0	Maintenance console	2 Missile area
8	0.03000	00.1	0	0	0	0	0	2	0	0	0	0	0	0	0	0	0	1	0	0	0	0	0	0	0	0	0	0	0	0	Launcher 1	
9	0.03000	00.1	0	0	0	0	0	2	0	0	0	0	0	0	0	0	0	1	0	0	0	2	0	0	0	0	0	0	0	0	Launcher 2	
10	0.14000	00.2	0	0	0	0	0	0	1	0	0	0	0	0	0	0	0	0	0	0	1	0	1	0	0	0	0	0	0	0	Hoist/handling equipment 1	
11	0.14000	00.2	0	0	0	0	0	0	1	0	0	0	0	0	0	0	0	0	0	0	1	0	1	0	0	0	0	0	0	0	Hoist/handling equipment 2	
12	0.14000	00.6	0	0	0	0	0	0	0	0	0	0	0	0	0	0	0	2	0	0	0	1	0	0	0	0	0	0	0	0	Portable test equipment	
13	0.14000	00.2	0	0	2	0	0	0	0	0	0	0	0	0	0	0	0	2	1	0	0	0	0	0	1	0	0	0	0	0	Reactor plant control console 1	3 Command and control area
14	0.14000	00.2	0	0	2	0	0	0	0	0	0	0	0	0	0	0	0	2	1	0	0	0	0	0	1	0	0	0	0	0	Reactor plant control console 2	
15	0.03000	01.0	0	0	0	0	0	1	0	0	0	0	0	1	1	0	0	2	0	0	0	0	0	0	0	0	1	0	0	0	Navigation/steering equipment	
16	0.01000	00.1	0	0	0	0	0	1	0	0	0	0	0	1	1	0	0	2	1	0	0	0	0	0	1	0	0	0	0	0	Missile fire control console 1	
17	0.01000	00.1	0	0	0	0	0	1	0	0	0	0	0	1	1	0	0	2	1	0	0	0	0	0	1	0	0	0	0	0	Missile fire control console 2	
18	0.03000	01.2	2	0	0	0	0	0	0	0	0	1	0	0	0	0	0	2	0	0	0	0	0	0	1	0	0	0	0	0	Radio/sonar equipment	
19	0.03000	02.0	0	0	0	0	0	0	0	0	0	0	0	0	0	0	0	2	0	0	0	0	0	0	1	0	0	0	0	0	Maintenance equipment	
20	0.01000	02.0	0	0	0	0	0	1	0	0	0	0	0	0	0	0	0	1	1	0	0	0	0	0	0	0	0	0	0	0	Environment monitor console	
21	0.30000	00.5	0	0	0	0	0	0	0	1	0	0	1	0	0	0	0	1	0	0	0	0	0	0	0	0	0	0	0	0	Medical electronics	4 Sick bay
22	0.00300	03.0	0	0	0	0	0	0	0	2	0	0	0	0	0	0	0	0	0	0	1	0	0	0	0	0	0	0	0	0	Mess equipment	5 Galley quarters

Table 6.3 Personnel Cross-training Data

	1	2	3	4	5	6	7	8	9	10	11	12	13	14	
1	0	0.30	0.05	0	0.30	0	0	0	0.1	0	0	0	0	0	Radar/sonar
2	0.20	0	0	0	0.25	0.05	0	0	0.03	0	0	0	0	0	Electronics technician
3	0.20	0.05	0	0	0.05	0	0	0.05	0.03	0	0	0	0	0	Nuclear reactor technician
4	0.05	0	0	0	0	0	0.10	0	0.03	0	0	0	0	0	Steam fitter
5	0.20	0.50	0	0	0	0	0	0.10	0.03	0	0	0	0	0	Electricians mate
6	0.10	0.30	0	0	0.30	0	0	0	0.03	0	0	0	0	0	Gunners mate
7	0.10	0.01	0.02	0.08	0.02	0.10	0	0	0.03	0	0	0	0	0	Boatswains mate
8	0.05	0	0	0	0	0	0	0	0.20	0	0	0	0	0	Pharmacists mate
9	0.05	0	0	0	0	0	0.05	0	0	0	0	0	0	0	Cook/baker
10	0	0	0	0	0	0	0	0	0	0	0	0.25	0.50	0	Electronics
11	0	0	0	0	0	0	0	0	0	0	0	0	0	0	Medical
12	0	0	0	0	0	0	0	0	0	0.60	0	0	0.80	0.10	Fire control
13	0	0	0	0	0	0	0	0	0	0.20	0	0.40	0	0.20	Staff
14	0	0	0	0	0	0	0	0	0	0	0	0.20	0.20	0	Power plant

PERSONNEL (line): rows 1–9
TYPE (staff): rows 10–14

Table 6.4   Parameters and Constants for Various Runs

Parameters		A	B	C	D
		Value, in Runs			

Parameters		A	B	C	D
Crew size increment	Δ	2	4	2	2
Morale threshold	M	0.400	0.400	0.400	0.400
Working hours/day	W	12.000	8.000	12.000	12.000
Probability of emergency	P	0.040	0.040	0.040	0.040
Iterations/run	N	2	2	2	2
Initial pseudo-random number (octal)	R_0	437275140113	—	—	—
Constants					
Average proficiency, primary specialty	$K_1{}^1$	0.750	0.750	0.500	0.500
σ Proficiency, primary specialty	$K_2{}^1$	0.100	0.100	0.100	0.100
Minimum proficiency, primary specialty	$K_3{}^1$	0.550	0.550	0.300	0.300
Maximum proficiency, primary specialty	$K_4{}^1$	1.000	1.000	0.750	0.750
Average proficiency, secondary specialty	$K_5{}^1$	0.500	0.500	0.400	0.400
σ Proficiency, secondary specialty	$K_6{}^1$	0.100	0.100	0.100	0.100
Minimum proficiency, secondary specialty	$K_7{}^1$	0.200	0.200	0.100	0.100
Maximum proficiency, secondary specialty	$K_8{}^1$	1.000	1.000	0.600	0.600
Probability of rank 1	$K_1{}^2$	0.500	0.500	0.500	0.500
Probability of ranks through 2	$K_2{}^2$	0.750	0.750	0.750	0.750
Probability of ranks through 3	$K_3{}^2$	0.880	0.880	0.880	0.880
Probability of ranks through 4	$K_4{}^2$	0.950	0.950	0.950	0.950
Probability of ranks through 5	$K_5{}^2$	0.980	0.980	0.980	0.980
Probability of ranks through 6	$K_6{}^2$	1.000	1.000	1.000	1.000
Probability of rates through 1	$K_{11}{}^2$	0.500	0.500	0.500	0.500
Probability of rates through 2	$K_{12}{}^2$	0.830	0.830	0.830	0.830
Probability of rates through 3	$K_{13}{}^2$	0.980	0.980	0.980	0.980
Probability of rates through 4	$K_{14}{}^2$	1.000	1.000	1.000	1.000
Weights for self orientation	$K_1{}^3$	0.250	0.250	0.250	0.800
Weights for crew orientation	$K_2{}^3$	0.400	0.400	0.400	0.100
Weights for mission orientation	$K_3{}^3$	0.350	0.350	0.350	0.100
Coefficient for total efficiency	$K_1{}^4$	1.000	1.000	1.000	1.000
Coefficient for E_{su}	$K_2{}^4$	0.350	0.350	0.350	0.350
Coefficient for E_{cu}	$K_3{}^4$	0.150	0.150	0.150	0.150
Coefficient for E_{pu}	$K_4{}^4$	0.300	0.300	0.300	0.300
Coefficient for E_{eu}	$K_5{}^4$	0.200	0.200	0.200	0.200
Probability of promotion/day	$K_1{}^5$	0.001	0.001	0.001	0.001
Probability of sickness/day	$K_2{}^5$	0.010	0.010	0.010	0.010
Average sickness duration	$K_3{}^5$	2.500	2.500	2.500	2.500
Filter constant	$K_1{}^6$	0.010	0.010	0.010	0.010
Filter constant	$K_2{}^6$	1.800	1.800	1.800	1.800
Filter constant	$K_3{}^6$	-0.810	-0.810	-0.810	-0.810
Constant of expectation	$K_1{}^7$	1.000	1.000	1.250	1.250

to the crew for each mission simulation run. Set B indicates an 8-hour day as standard and since more men will be required to accomplish the same work without overtime, the crew size was increased by four men on each mission simulation run. Set C forced the initial proficiency level of the crew to be substantially reduced from those given for set A.

Set D represents an experiment with a highly self-oriented crew as opposed to the relatively evenly balanced, slightly crew-oriented crew of set A. Runs C and D also contained a higher value for K_1^r, thus creating a higher threshold for acceptable work. In these runs the level of working efficiency acceptable for a successful action unit performance was effectively increased 25 percent.

MISSION DATA

The crew "aboard" this hypothetical submerged system had as its major goals to cruise to their destination, to remain on underwater alert, and to maintain their proficiency by training. To implement this major mission three types of workdays were defined as follows:

Type number	Type of day
1	cruise
2	stationary
3	stationary with emergency drills

The sequence of operations to be accomplished in each of these three day types is shown in Table 6.5. These sequences are composed of action units representing engine watches, preventive maintenance, mess, sick call, simulated missle launch, and the like. For each action unit data are supplied for the average execution time, action unit orientation, action unit type, equipments required, and other action unit data specified in the mission data.

RUNS PERFORMED

Employing the above data, computer simulations were performed to assess the effects of the variations introduced by modifying initial proficiency, workday length, orientation, the threshold for acceptable work, and crew size on the output of the model. The results from these runs are presented below. A discussion of the rationale of the results in relationship to the logic of the model is also presented.

Table 6.5 Operations Accomplished in Each Type of Day

1	2345	6	7	8	9	10	11	
4.0	1011	0.100	0.200	0.800	1	0	0110000000001110010000	
4.0	1001	1.000	1.000	1.000	1	0	0000000000000000000100	
2.0	0101	0.200	0.800	0.200	1	0	0000000000000000000001	
4.0	1011	0.100	0.200	0.800	1	1	0110000000001110010000	
4.0	1001	1.000	1.000	1.000	1	2	0000000000000000000100	
2.0	0101	0.200	0.800	0.200	1	1	0000000000000000000001	
2.0	0102	0.100	0.750	0.200	3	4	0000000000000000000010	
4.0	1011	0.100	0.200	0.800	1	4	0110000000001110010000	Cruise
4.0	1001	1.000	1.000	1.000	1	5	0000000000000000000100	
2.0	0102	0.100	0.100	0.600	1	8	1111110000000000000000	
2.0	0104	0.100	0.800	0.200	1	7	0000000000000000000010	
2.0	0101	0.200	0.800	0.200	1	9	0000000000000000000001	
4.0	1011	0.100	0.200	0.800	1	8	0110000000000110010000	
4.0	1001	1.000	1.000	1.000	1	9	0000000000000000000100	
2.0	0101	0.200	0.800	0.200	1	12	0000000000000000000001	
1.0	0102	0.100	0.100	0.600	1	8	0000000111100000000000	
4.0	1011	0.100	0.200	0.800	1	13	0110000000001110010000	
4.0	1001	1.000	1.000	1.000	1	14	0000000000000000000100	
2.0	0101	0.200	0.800	0.200	1	15	0000000000000000000001	
4.0	1014	0.100	0.200	0.800	1	17	0110000000000110010000	
4.0	1004	1.000	1.000	1.000	1	18	0000000000000000000100	
4.0	1001	1.000	1.000	1.000	1	0	0000000000001100000100	
4.0	0101	0.200	0.800	0.200	1	0	0000000000000000000001	
4.0	1001	1.000	1.000	1.000	1	1	0000000000001100000100	
2.0	0101	0.200	0.800	0.200	1	2	0000000000000000000001	
2.0	0102	0.100	0.750	0.200	3	3	0000000000000000000010	
4.0	1001	1.000	1.000	1.000	1	3	0000000000001100000100	
0.3	0111	0.200	0.200	1.000	3	0	000000011110011100000	
0.3	0111	0.200	0.200	1.000	3	0	000000011110011100000	Stationary
2.0	0101	0.200	0.800	0.200	1	4	0000000000000000000001	
1.0	0102	0.100	0.100	0.600	1	6	0000000000001111110100	
4.0	1001	1.000	1.000	1.000	1	6	0000000000001100000100	
2.0	0104	0.100	0.800	0.200	1	7	0000000000000000000010	
2.0	0101	0.200	0.800	0.200	1	9	0000000000000000000001	
1.0	0102	0.100	0.100	0.600	1	6	1111110000000000000000	
0.4	0111	0.200	0.300	1.000	2	8	000000011110011100000	
4.0	1001	1.000	1.000	1.000	1	11	0000000000001100000100	
2.0	0101	0.200	0.800	0.200	1	13	0000000000000000000001	
1.0	0102	0.100	0.100	0.600	1	13	0000000111100000000000	
4.0	1004	1.000	1.000	1.000	2	16	0000000000001100000100	
4.0	1001	1.000	1.000	1.000	1	0	0000000000001100000100	
2.0	0101	0.200	0.800	0.200	1	0	0000000000000000000001	
4.0	1001	1.000	1.000	1.000	1	1	0000000000001100000100	
2.0	0101	0.200	0.800	0.200	1	2	0000000000000000000001	
2.0	0102	0.100	0.750	0.200	3	3	0000000000000000000010	
4.0	1001	1.000	1.000	1.000	1	3	0000000000001100000100	
0.3	0111	0.200	0.200	1.000	3	0	000000011110011100000	
0.3	0111	0.200	0.200	1.000	3	0	000000011110011100000	
2.0	0101	0.200	0.800	0.200	1	4	0000000000000000000001	Stationary with Emergency
1.0	0102	0.100	0.100	0.600	1	6	0000000000001111110100	Drills
4.0	1001	1.000	1.000	1.000	1	6	0000000000001100000100	
2.0	0104	0.100	0.800	0.200	1	7	0000000000000000000010	
2.0	0101	0.200	0.800	0.200	1	9	0000000000000000000001	
1.0	0102	0.100	0.100	0.600	1	6	1111110000000000000000	
0.4	0111	0.200	0.300	1.000	2	8	000000011110011100000	
4.0	1001	1.000	1.000	1.000	1	11	0000000000001100000100	
2.0	0101	0.200	0.800	0.200	1	13	0000000000000000000001	
1.0	0102	0.100	0.100	0.600	1	13	0000000111100000000000	
4.0	1004	1.000	1.000	1.000	2	16	0000000000001100000100	
1.0	0112	0.800	0.500	0.800	2	15	1000100000000000010010	

Key
1. Average execution time in tenths of hours
2. Fixed time indicator
3. Repetition code
4. Communication indicator
5. Carry over code
6. Self orientation of action unit
7. Crew orientation of action unit
8. Mission orientation of action unit
9. Action unit type code
10. Prior action unit number
11. Equipment systems required

115

CREW EFFICIENCY

The efficiency of the simulated crews operating in the four series of runs is presented in Figure 6.1. No criterion data are available against which to verify the absolute value of the results; however, these efficiency values appear intuitively reasonable. Relatively high end-of-mission efficiency values, 0.80 and 0.79, were achieved by crews on runs A and B, the two runs involving crews that possessed the highest initial proficiencies. Since the only difference in initial conditions between runs A and B was the number of working hours per day (12 and 8, respectively), it can be seen that only about a one-percent decrease in efficiency was attributable to a reduction in that parameter from a 12-hour day to an 8-hour day. Such results represent an example of the potential uses to which the model can be put.

Focusing attention on run C we note a substantial efficiency reduction (almost 7 percent in comparison with run A). This decrease is attributable to the reduction in average initial prime proficiency from $K_1^1 = 0.75$ to $K_1^1 = 0.50$ for the average crew member. The effects of the increase in the performance expectation constant K_1^7 from 1.0 to 1.25 are discussed later. In accord with expectation each of these results seems

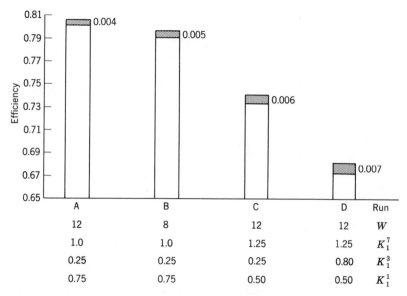

Figure 6.1   Final efficiency and efficiency increase.

logical and appears to support a contention that the model is sensitive to variation in the initial proficiency of the simulated crew members.

The lowest efficiency was indicated for run D—a further reduction of about 6 percent from the crew efficiency of run C. This lowering is attributable solely to the increase in the initial value of self-orientation for the average crew member from a more normal average self-orientation value of 0.25, as selected for runs A, B, and C, to an intentionally excessive value of 0.80 in run D. Logically one would not anticipate that a highly "self-centered" crew would perform efficiently as a team.

EFFICIENCY INCREASES

Figure 6.1 also indicates that increases in efficiency were achieved during the 10-day mission. These increases are represented by the striated areas of the figure. As would be expected for short missions a relatively small efficiency increase (from 0.4 percent to 0.7 percent) was indicated. Moreover the smaller increases were noted on the runs A and B, in which the initial proficiency itself was higher, and larger increases are noted for runs C and D, in which lower initial proficiency values are indicated. This result also seems reasonable.

The smoothed efficiency value $\bar{E}_d$ calculated in such a way as to damp out "chance" fluctuations is based on E_d and on the smoothed values $\bar{E}_{d-1}$ and $\bar{E}_{d-2}$ for the previous two days. Figure 6.2 presents a graphic illustration of both of these effects.

The raw, daily efficiency values show in accordance with logic a sharp initial rate of efficiency increase and realistic, random daily variation. The lower curve of Figure 6.2 presents the effects of the smoothing function on the calculated daily crew efficiency for the same sample iteration. It will be noted that the smoothing appears to be relatively heavy. This is considered desirable in a model in which the computation of many of the variables is based on random but constrained effects. Moreover it appears from the upper curve of Figure 6.2 that more "learning" took place during the first few simulated mission days—an entirely reasonable effect.

PRODUCTIVE TIME

The previously discussed results suggested that a 12-hour workday would not result in a crew efficiency for the mission simulated that was lower than an 8-hour workday. Let us now consider the results relating to the productive time of the crew members during the 10-day mission. Figure 6.3 presents average hours worked per day as a function

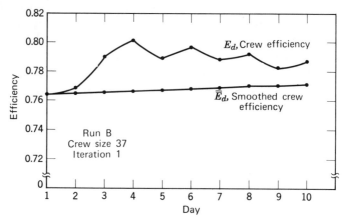

Figure 6.2 Effects of efficiency smoothing.

of crew size with specific reference to the 12- and the 8-hour workdays. Here two major effects are indicated. The first effect is a slight reduction in the average working hours as crew size is increased. A rough estimate from run B results indicates that for each new man added to the crew the thirty to forty others worked about 0.05 hour (3 min) less per day. Thus the added man worked about 6.5 hours while the others together worked a total of only 2 hours less.

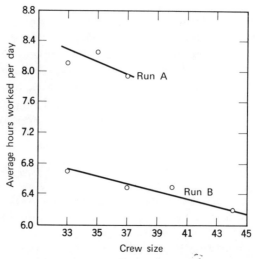

Figure 6.3 Average hours worked per day as a function of crew size.

In run A, in which the limit was 12 hours work per day before calling overtime, about 8.2 hours, on the average, was actually spent in work. It is noted, however, that the reduction in the working day from 12 to 8 hours reduced the average number of hours worked by only 1.7 hours on the average—an important result of the model. The fact that such a large percentage of time in the mission workday remains un-worked, on the average, results in a large measure from the restricted abilities of several crew members whose specialities were infrequently called for by the mission and who were infrequently cross-trained for other tasks. These men are consequently lightly loaded during the mis-sion. In fact a major finding can result from these and similar runs with respect to this crew idle time. Run A crew members are under-worked either because of mission imbalance (insufficient need for that specialty) or by cross-training imbalance (lack of ability to fill available time by performing required work in an alternate specialty). Stated alternatively the simulated run A crews, although putting in a long workday, were not overworked.

By this simple illustration it is noted that the model possesses the potential for making contributions to general systems effectiveness studies by providing data on the crew and mission balance. Much of the imbalance in this test mission could have been avoided by higher cross-training probabilities for the men whose workday was not filled. The mission clearly does not present a full workday for these men.

More specifically investigation of summarized daily printouts indicated that an imbalance of needs existed from one type of day to another type. For example, on cruise days there was almost no work assignable to electronics technicians and gunners mates; on the other hand, on sta-tionary days the computer did not assign more than a few hours work per day to radar-sonar operators, steam fitters, electronics and staff officers in their primary specialities. Alternate personnel planning is clearly indicated as desirable by even this limited number of runs.

WORK COMPLETED

Figure 6.4 presents the total number of action units successfully com-pleted by run as a function of crew size. In run B fewer action units were performed; that is, as might be expected a greater number of action units were ignored in order to avoid overtime than in the other runs for which $W = 12$. This difference amounted to about one to two action units per day on the average, depending on crew size (the work-ing time for the average action unit over the mix of days in the 10-day mission was 2.7 hours). The slope of the line for run B shows that

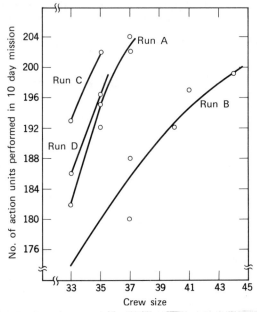

Figure 6.4   Number of action units performed as a function of crew size.

addition of a single crew member resulted in accomplishment of only about two or three additional action units during the entire 10-day mission (equivalent to about 5 to 8 hours of work). This figure is raised to about five additional action units accomplished (or about 13.5 hours worked per mission) per crew member for runs A and D—that is, with a 12-hour workday.

COMPLETION OF NONESSENTIAL ACTION UNITS

The summarized results regarding performance of nonessential action units are shown in Figure 6.5. The logically expected trend showing performance of more nonessential tasks as the crew size and working hours increased is clear. The average difference between runs A and B is about nine action units—or the equivalent of about one action unit per mission day—even though the workday was 4 hours longer. (The average predicted completion time of action units from input data on cruise days was 3.1 hours and for stationary days the average was 2.1 hours.) The leveling off effect for performing additional nonessential action units as crew size increases is also demonstrated for run B.

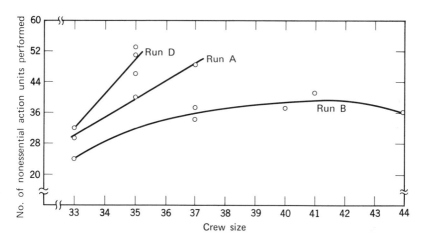

Figure 6.5 Number of nonessential action units performed as a function of crew size.

ACTION UNIT FAILURES

In the following discussion of the number of action unit failures an action unit is considered poorly performed or failed if the required efficiency level is not attained on initial performance or on repetition. Thus an action unit on which initial performance is unsatisfactory but which is repeated satisfactorily ("touched up") is not considered to be failed.

Figure 6.6 summarizes the total number of action unit failures in the average 10-day mission for each of the four runs. The results indicate that for the shorter workday (run B as compared with run A) on which fewer action units were attempted, fewer failures also occurred. The further reduction in run C is the result of two counterbalancing effects. First, the increase in the constant of expectation makes performance satisfactory only if 25 percent greater group efficiency is achieved than for runs A and B, all other things being equal. This effect would, by itself, result in a greater number of failures in B. The effect, however, is counterbalanced by the reduction in the average initial proficiences of crew members. For run C the starting proficiency was 0.5 as opposed to 0.75 for runs A and B. (Similarly proficiencies in the alternate specialties were reduced from 0.5 to 0.4). We note that the proficiency of the performing group enters into both sides of the failure-success equation. However, the work group's proficiency is only one of five contribut-

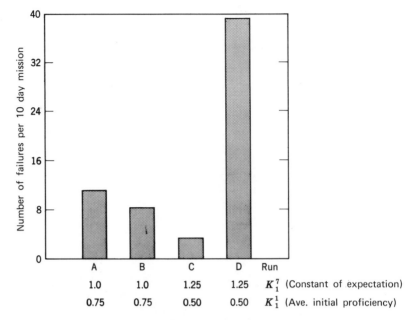

Figure 6.6   Total action unit failures per mission.

ing terms in the calculation of E_u on the left-hand member of the inequality, while it constitutes the only variable on the right-hand side. Thus decreases in the group proficiency (highly dependent on average initial proficiency constants) predominate over the effect of a change in the constant of expectation resulting, as expected, in fewer failures. In humanistic terms this effect may be viewed as the equivalent of setting an aspiration level that is higher than a previous attainment level and the high aspiration level providing the motivation for achieving the aspiration level.

In Figure 6.6 a significant reversal is shown for run D. This reversal indicates a strong relationship between the increase in self-orientation and action unit failures. This greater number of failures is considered a logical result of a highly self-oriented crew and is readily explained as a result of the logic of the model. With the average crew member having a self-orientation of 0.8, the self-orientation of the work group will be even higher because the method of calculating self-orientation of the crew emphasizes the largest orientational component. Since most of the action units of the 10-day mission are crew or mission oriented and the group and crew are self-oriented, we are presented with an orientation that is predominantly case E as described in Chapter III. This case produces very low values of psychosocial efficiency which, in turn, is the most dominant term in the calculation of action unit efficiency. Thus substantial reductions in success percentages—that is, more failures—are to be expected. For comparison, values for E_{su} in the other runs were predominantly in the range 0.5–0.9.

Figure 6.7 presents by run the effect of crew size on action unit failures. An increase of about one action unit failure per mission with the addition of each crew member is shown for run B. This small increase is probably a result of the increase in the number of extra, nonessential action units accomplished by larger crews. An interesting result in Figure 6.7 is that for run D. In this case, in which the simulated crews were highly self-oriented, the addition of a single man to the crew increased the number of failures by roughly a full order of magnitude greater than for more crew and mission oriented crews. Thus the model's results suggest that the addition of a self-oriented person to a highly self-oriented crew will not serve a useful purpose in terms of mission accomplishment. This result also seems quite reasonable and in accord with group dynamic theory.

CREW MORALE

Figures 6.8 and 6.9 display the effects of the number of working hours per day and failure percentages on the "morale" of the crew.

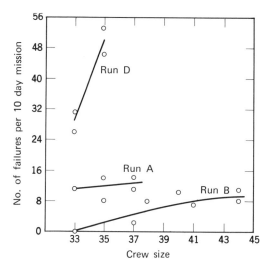

Figure 6.7   Number of failures as a function of crew size.

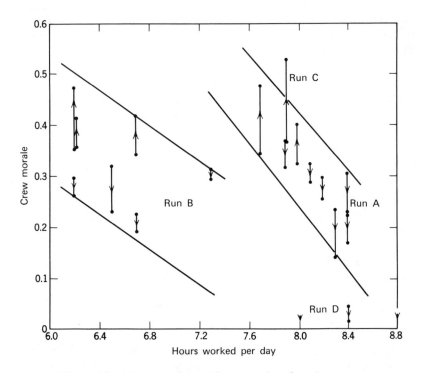

Figure 6.8. Crew morale as a function of working hours.

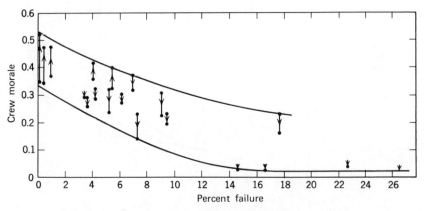

Figure 6.9   Crew morale as a function of percentage failures.

In these figures the vertical line segments connect points that represent the value of crew morale at the beginning and at the end of the mission. The arrows indicate direction of change of the morale variable. Figure 6.8 indicates the expected relationship; that is, crew morale bears an inverse relationship to the average number of hours worked per day.

The composite results for all runs (Figure 6.9) also clearly indicate the logically expected reduction in crew morale as the percentage of work that results in failure is increased. The crew morale "band," about 0.2 morale units wide, shows a decrease of about 0.025 units for each 1-percent increase in failures up to about 10-percent failures. Thereafter the band appears to approach zero morale asymptotically. The break in the direction of morale change in the two figures, at around a crew morale value of 0.35, is believed to be an artifact of the morale deviation factor incorporated in the model. Evidently simulated crews starting with low morale values tend to perform less proficiently and the result is a lowering of morale as defined and calculated within the model. This result is elaborated on the following section.

COHESIVENESS AND ORIENTATIONS

The model's calculations yield two cohesiveness indices. The first cohesiveness index $I_c{}^1$ based on the variance of the crew's orientation value is determined to be the largest of the three (crew, mission, and self-orientation) and is not discussed here. In the second index $I_c{}^2$ the largest variance of the three orientational values is employed as the basis for the calculation. The resulting second index $I_c{}^2$ presented a pattern that is rationally explained in terms of expected performance (Figure 6.10). Here we see generally poorer cohesiveness among crews

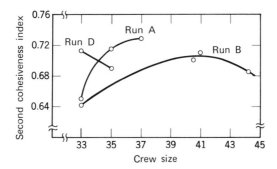

Figure 6.10   Second cohesiveness index as a function of crew size.

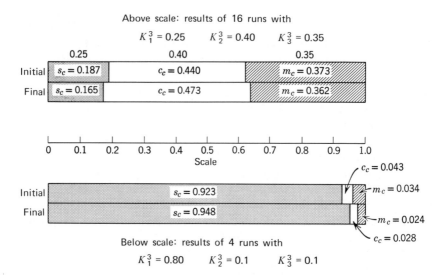

Above scale: results of 16 runs with

$K_1^3 = 0.25$ $K_2^3 = 0.40$ $K_3^3 = 0.35$

Below scale: results of 4 runs with

$K_1^3 = 0.80$ $K_2^3 = 0.1$ $K_3^3 = 0.1$

Figure 6.11   Initial/final crew orientation.

who worked shorter hours (run B) as compared with the 12-hour days of run A. An interesting effect clearly noticed for run B (and probably run A if sufficient data were available) is that cohesiveness reached a peak value as a function of crew size. Here a 10-percent increase in cohesiveness was indicated by adding 8 men to the minimum 33-man crew; further additions yield reductions in cohesiveness as calculated and defined.

In run D, with its highly self-oriented crew members, a rapidly declining cohesiveness was indicated in the region studied. It appears as though the segment shown is on the declining portion of the inverted U curve. This decline suggests that for highly self-oriented crews improved cohesiveness can be achieved only when such small crews are used that too little work would be done or that large self-oriented groups are not considered to be cohesive by the model.

Figure 6.11 summarizes the changes in crew-orientation values accumulated over all runs. The top portion of the figures shows that when corresponding end-of-mission orientation values for the crews on all simulated missions in runs A, B, and C were averaged, crew orientation increased at the expense of mission and self-orientation, both of which showed small decreases. The average value of c_c at the start of the mission was 0.440 and this increased to the dominant value of 0.473. The lower portion of the figure shows the results from run D in which self-orientation was predominant. The average values of initial orientation do not match the corresponding K^3 values because crew orientations are calculated by normalizing the sums of the squares of individual crew-member orientations rather than by simple summation. In this run the self-orientation of the crew increased slightly over the mission orientation.

SICKNESS

The effect of crew-member illness was simulated by setting the appropriate constants $K_2{}^5$ and $K_3{}^5$ so that each crew member was "sick," on the average, once every 100 days for a period averaging 2.5 days. As the number of sick days encountered during the mission increased the average hours worked by the remaining crew members also tended to increase (Figure 6.12).

OVERTIME

Even though relatively high average idle time figures were indicated as model results from the mission, certain crew members worked overtime during the simulated mission. The average overtime hours per crew

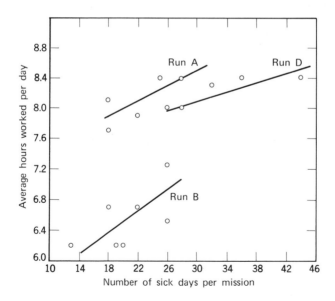

Figure 6.12 Average hours worked as a function of number of sick days.

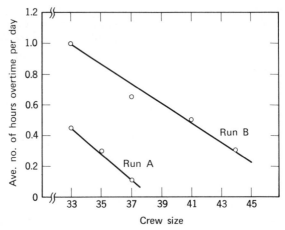

Figure 6.13   Average hours overtime as a function of crew size.

member per day are plotted for runs A and B in Figure 6.13, which suggests that the simulated reduction of the nominal working day from 12 to 8 hours resulted in about a 36-min per day reduction in overtime spent by the average crew member. The addition of one man to the crew caused a reduction in overtime per man by 38 min per day for run B, and 52 min per day for run A. This is equivalent to a reduction of 25 hours of overtime per day for a 40-man crew for run B. To avoid almost all overtime a 37-man crew with a 12-hour workday or more than 45 men with an 8-hour workday is indicated.

PREDICTIVE VALIDITY SIMULATION

The results of a test of the predictive validity of the group model are now presented. In this test only data available to system design personnel early in the system planning stage were employed as the input data to the model. Then the model was run through a typical mission and the model's results were compared with actual mission data.

MISSION AND SYSTEM SIMULATED

The specific mission selected for a test application for model validation study was a realistic 21-day training mission of a nuclear FBM submarine (SSBN). This complex, mobile weapons system has for its primary mission the delivery of retaliatory ballistic missile attacks on assigned strategic targets. (Secondary missions of self-defense, tactical attack, and the like, while within the capability of the FBM submarine, are subordinate to the primary mission.) The FBM submarine driven by steam turbines powered by water-cooled nuclear reactors is capable of almost unlimited submerged operations over that 70 percent of the earth's surface covered· by the oceans. Each FBM submarine carries Polaris missiles, which have sufficient range to give the FBM/Polaris an ability to reach any point on the earth's surface from international waters. From the time it leaves home port until it returns there is no need for the FBM submarine to surface, snorkle, or raise its periscope under normal circumstances. Environmental control, including the manufacture of oxygen from sea water, is completely self-contained within the weapons system. An inertial navigation system provides accurate ship's position information without need to surface. Finally, world-wide radio communication with other submarines without compromising location is possible. Thus it is apparent that one of the principal limitations in the operation of the FBM nuclear submarine is the endurance of the crew—for missions of long duration crew performance is unquestionably the primary limitation.

THE MISSION

The patrol on which no emergency situation arises is essentially a training mission consisting of three types of activities: (1) individual and small-group training on routine assignments, (2) larger team training on emergency procedures, involving less than the whole crew, and (3) total-crew training in the implementation of the primary mission. Therefore three types of "mission days" were developed and analyzed numerically for use in the model. The first consisted of a reiterative cycle of six identical 4-hour watches during which all the simulated equipments were portrayed as being manned for routine operations and maintenance functions of their compartments. In the second type a half-hour general quarters drill was included and in the third type emergency drills and training assignments for personnel in selected compartments were also performed.

A 21-day mission was determined that was composed of the three types of mission days.

INPUT DATA

In order to derive the mission input data, the required intercommunication matrix, and certain required parameters and constants *Personnel*

Research Memos, Personnel Research Notes, Operator Sequence Analyses, Organization and Manning Studies, and similar documents published while the system was in its design and development phases were examined.

Data were prepared for the simulation of thirteen major equipments in six stations (the ship control center, the missile control center, the missile compartment, the torpedo room, radio control, and the sonar control room). Personnel types were selected to be consonant with seven Naval enlisted rates and six officer ranks. About 70 action units were identified for each of the 21 days of the simulated mission and about 1900 action units were simulated in a mission, including repairs and action unit repeats.

OBJECT OF TESTS

The primary purpose of performing these simulations was to validate the group simulation model. By validation is meant the determination of the ability of the model to predict empirical, independently derived, outside criterion data. Accordingly a comparison of predictions made on the basis of data available during the design and development with operational data was selected to represent one test of the ability of the technique to yield usable operational predictions. The results may be classified in three ways as follows:

1. Those results for which objective criterion data were available from U.S. Navy sources—primarily manning data.
2. Those results for which no outside criteria data were available in records but for which estimates could be obtained by interviews with crew members—this includes the performance effectiveness and motivation-morale predictions.
3. Those results that were not amenable to outside verification through records or interview methods but that were reviewed from the point of view of their coherency or their agreement with logical expectations.

CREW COMPOSITION DETERMINATION

The first object of the FBM submarine simulation was to compare the computer model's prediction of the required crew composition, given the mission and other input data, with the actual manning found feasible in operation. At the outset an initial computer simulation was performed in order to allow the computer to make an initial estimate of required manning. Because the computer's initial manning estimate is based on an impoverished situation, this initial simulation yielded a number of situations of undermanning. Subsequent runs with increasing crew sizes were then completed to reduce the average overtime hours worked by crew members. This was done on a selective basis by type of personnel (without using the Δ feature). These runs successively involved 50, 55, 58, and 61 crew members for the compartments simulated. The 61-man crew was simulated for both an 8-hour and a 12-hour working day.

The computer predictions on the basis of engineering design data of the required manning for the compartments involved and the actual current manning are presented in Figure 6.14, which shows a close correspondence between the actual manning and the computer model's predictions of current required manning. The computer tended to be conservative and to underpredict the number of missile technicians, quartermasters, and sonar technicians required by one man. Similarly the computer underpredicted the number of seamen by four and the number of torpedomen by seven.

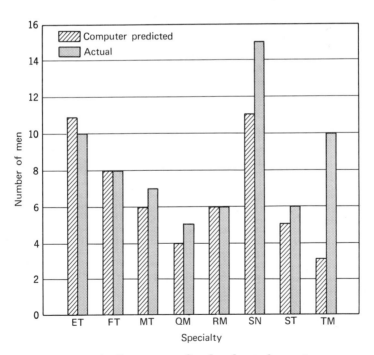

Figure 6.14 Computer predicted and actual manning.

The number of men required to man the navigation, operation, and weapons departments was predicted exactly for two of the eight ratings involved and if a difference of plus or minus one man is allowed, eight of the ten manning predictions were acceptable. The error of the prediction in regard to the torpedoman rating was explained on the basis of faulty action unit analytic computer input data, as was the underprediction for seamen. Moreover the underprediction for seamen is not considered to be serious since when the torpedoman's estimate was ignored, eleven of the twelve respondents to the interviews (described below) thought that they would "do as well" or "do better" with the computer selected manning than with the current manning. Even with the relatively poor predicted manning for torpedomen, eight of the 12 experienced FBM crew members interviewed thought that they would "do as well" or "do better" with the computer predicted manning as compared with the current manning. When the torpedomen were not considered, three of those who replied "do worse" indicated that they believed that they would do "as well" with the computer generated manning as with the current manning. The product moment correlation between the computer generated mannings and the actual current manning was 0.89 when torpedomen were eliminated and was 0.98 when both seamen and torpedomen were eliminated from consideration. With the exception of the torpedoman prediction these data are believed to indicate adequate effectiveness for the computer generated predictions as supported by the interview findings.

CRITERION DEVELOPMENT

Since records or reports were not available in the areas of performance effectiveness and motivation-moral, the criterion data in these two areas were based on the results of interviews with officers and chief petty officers from the navigation department, the operations department, and the weapons department of current FBM submarines. A semistructured interview was developed that allowed collection of the required data. The interview in its final form was administered to a total of twelve officers and chief petty officers from each of four ships of a submarine flotilla. Thus three crew members, each representing a different department of each ship, were individually interviewed. Each interview consumed about 1 hour, allowing sufficient time for thorough explanation of the meanings of questions and for probing and exploring the nuances of the responses of the crew member interviewed.

The following comparisons of computer and interview results are made on the basis of a 61-man crew with an 8-hour workday.

Table 6.6   Some Interview Results

	Very Low	Slightly Low	About Right	Slightly High	Very High	No Response
Average crew proficiency estimate	0	1	9	2	0	0
Average satisfaction of supervisor estimate	2	1	9	0	0	0
Number of nonessential action units skipped	0	1	5	5	0	1

CREW PROFICIENCY

For the computer selected manning described above the computer model yielded an average crew proficiency estimate of 0.744 at the end of the 21-day mission. The twelve respondents were asked to rate this crew proficiency estimate along a five-point, Likert-type scale which ranged from "very low" through "about right" to "very high." The results are presented in the first row of Table 6.6 and indicate that nine of the twelve respondents thought the computer prediction to be "about right." Two of the remaining three believed the computer prediction to be "slightly high" and one respondent believed the computer calculation to be "slightly low." These results are interpreted to indicate that the simulation was effective in predicting the proficiency of the crew, at least in the opinion of the supervisors included in the sample.

CREW EFFICIENCY

Another set of data yielded by the computer model concerns the efficiency of the crew's work during each day of the simulated mission and indicates how well the crew performed. The respondents were asked to make a magnitude estimation along a continuous scale from 0 to 100 of the efficiency of the performance on a normal watch-standing task of a representative team in the compartment with which they were most familiar. Separate judgments were made for performance "near the beginning," "near the middle," and "toward the end" of a patrol. A wider divergence of opinion was elicited in the "near the beginning" estimates than for the others. This is believed to reflect the difference of opinion of the effects of the "upkeep" period that precedes a patrol. The mean values so elicited for each period were as follows: beginning = 69.1, middle = 79.7, and end = 85.6. The mean across these three values is 78.2, a value in close accordance with the average value

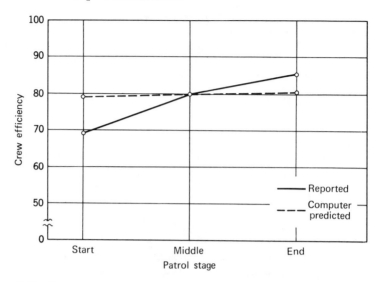

Figure 6.15 Computer predicted and reported effectiveness at various patrol stages.

of 80 yielded by the model. The mean estimates of the crew members and the value yielded by the computer for the various patrol stages are presented in Figure 6.15. In preparing Figure 6.15 the computer's estimate of the crew effectiveness on the fourth, the twelfth, and the nineteenth days of the computer simulated 21-day mission was employed. The chi square test, employed to test the statistical significance of the difference between the two sets of points represented in Figure 6.15, yielded a value of 0.0064 which with 4 degrees of freedom is not statistically significant. The reason for the overestimation for the computed effectiveness as compared with the reported effectiveness for the "near the beginning" comparison is believed to be a function of the depressed effectiveness mean brought about by a number of the respondents who reported very low effectiveness values at the very early stages of a patrol (two estimates of 50, one of 45, and one of 30).

ACTION UNITS PERFORMED UNSATISFACTORILY

The model predicted that the proportion of action units to be performed unsatisfactorily would be slightly less than 1 percent; that is, less than 1 percent of the action units would be performed below the expectation level of the supervisors. The supervisory personnel interviewed were also asked to rate this calculation of less than 1 percent on a 5-point scale. These respondents were told that if they thought the actual value was between 0 and 5 percent, an "about right" rating would probably apply. Nine of the twelve respondents indicated the predicted value to be "about right." (See Table 6.6)

HOURS WORKED

The interviewees were asked to estimate the number of hours greater than 8 that a crew member works each day. The mean number of "overtime" hours so elicited was 3 (an 11-hour workday) for the average FBM crew member. This workday includes in addition to normal watch standing (9 hours per day) housekeeping, qualification, and general ship maintenance. If the computer's estimate of average hours overtime work is corrected for 2 hours per day known to be spent on these general duties that were not included in the simulation, the simulation prediction of average overtime becomes 2.3 hours. Although somewhat conservative, this prediction is considered to be within a reasonable and acceptable error range and adequate for the early system planning purposes for which the simulation model is designed.

The crew members interviewed were shown a listing of the ratings involved in the computer simulation and were asked to indicate which, if any, customarily work overtime. The concensus of their responses was in agreement with the predictions of the computer model which (if qualification, ship maintenance, housekeeping, and related time is considered) indicated overtime for all of the ratings involved.

Similarly the computer simulation yielded a prediction of manpower shortage hours (execution time on action units for which more than one man is required but which are performed with reduced manning). For the 21-day mission and a 61-man crew the computer yielded a prediction of 17.3 hours manpower shortage per day or 3.7 percent. The interviewees were asked, "At your watch station, how often (in 10-percent intervals) is the manning for a task inadequate?"

Eight of the twelve interviewees responded "never" or "between 0 and 10 percent of the time." This value seems in adequate correspondence with the computer simulation result of 3.7-percent manpower shortage hours for the FBM system.

IGNORED OR POSTPONED ACTION UNITS

The experienced FBM officers and chief petty officers interviewed were asked to estimate the number of assigned but unessential tasks that are either postponed or ignored on a patrol. The mean of the twelve estimates was 32 percent—that is, about one third of the nonessential tasks being postoned or ignored. The simulation values for nonessential tasks skipped varied from day to day of the simulated patrol in accordance with situational requirements. The lowest percentage of nonessential tasks skipped on a given day was 11 percent and the highest was 45 percent. Over the 21-day simulated mission the average number of nonessential action units postponed or ignored was 20 percent or 12 percent under the mean of the interview sample. The fact that the simulation did not include housekeeping and similar tasks may account for the difference. As a check on the quantitative data the interviewees were advised of the computer estimate of the number of nonessential action units skipped and were asked to judge the adequacy of this estimate on a 5-point scale. A review of Table 6.6 implies that the qualitative judgments of the crew members interviewed do not agree with their quantitative estimates. When asked to make a quantitative estimate, the interviewees reported about one third of the nonessential tasks to be skipped. Yet when shown a lower computer estimate, the respondents felt that the computer estimate was "about right" or "slightly high." If one attempts to synthesize these disparate findings and to form a meaningful nexus with the computer estimate, it seems that the computer estimate, which falls between the two sets of judgments, is probably reasonably close to the actual value.

The computer model does not allow for the skipping and postponing of "essential" action units. As a check on this logic the respondents were asked to state the percentage of "essential" action units that are customarily completed on a patrol. The frequency distribution of replies was 100 percent = 6, 99 percent = 2, 98 percent = 1, 95 percent = 1, 90 percent = 2. If we consider the 95-percent responses or above to be an artifact of an unwillingness of some persons to reply to such a question in all-or-none terms, then ten of the twelve respondents can be said to agree with this aspect of the model's logic.

ORIENTATION

The crew members interviewed were asked to provide a magnitude estimation for the value of each of the orientations at the start and near the end of a patrol for a "typical" crew member and to indicate the direction of any orientational changes within the mission. The consensus seemed to be that crew members tend to become more self-oriented and less mission or patrol oriented as a patrol continues. Crew orientation was generally believed to remain about constant over the patrol. Comparison of these direction changes with those yielded by the computer simulation indicated little, if any, agreement. The computer model indicated that the trend includes an increase in mission orientation and decreases in both the crew and the self-orientations of the crew members. Analysis of the individual data of the interviewees indicated that about 25 percent of the interviewees would tend to agree with the computer model's directional predictions for self-orientation and crew orientation, while about 33 percent of them would agree with the directional prediction for mission orientation. These findings seem to suggest that the model's orientation results per se are marginal and that these variables should be considered solely as internal model elements used in the generation of other external model results.

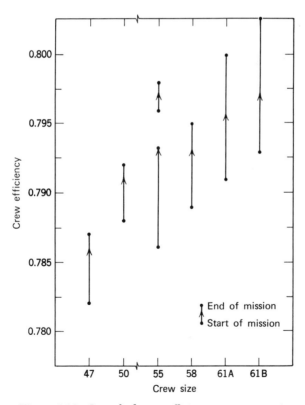

Figure 6.16 Smoothed crew efficiency versus crew size.

CREW EFFICIENCY

A number of other directional trend-data were available as the result of the parametric variation introduced into the simulation runs. Additionally certain directional tendencies, which were not amenable to check either through records or interview methods, were indicated within the various simulation runs.

Figure 6.16 displays the generated, smoothed crew-efficiency results as a function of crew size. For these runs larger crews tended to have higher crew efficiencies and to experience greater increases in efficiency during the mission. This finding seems logical since seriously undermanned ships would not be anticipated to perform with the greatest possible efficiency. To provide some concept as to the extent of the smoothing effect introduced into the calculation.

Figure 6.17 presents a plot of the smoothed and unsmoothed daily crew-efficiency values for the 21-day mission using the 61-man crew (12-hour day) run as an example. As indicated in Figure 6.17 rather wide day-to-day variations in unsmoothed efficiency occurred. The direction of the unsmoothed trend is, however, toward an increase in efficiency over the

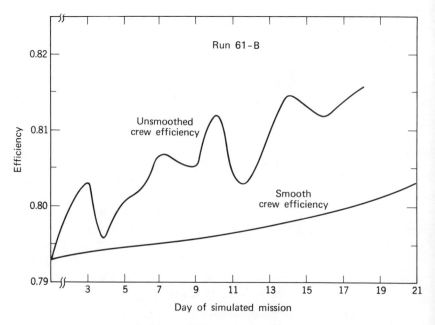

Figure 6.17   Effects of smoothing.

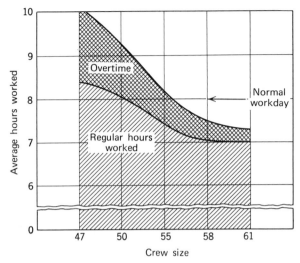

Figure 6.18   Average hours worked (regular versus overtime).

mission and even after 21 days, the efficiency of the crew was still increasing. The smoothing seems to have achieved the desired effect of reflecting consistent day-to-day values without variations that might tend to become confusing and be based on an artifact within the simulation of any single day's events.

PRODUCTIVE TIME

Figure 6.18 shows the model's output with respect to the average hours worked by the crew for all runs having an 8-hour workday. This average covers all crew members over the entire mission. As was expected increased numbers of personnel available generated, on the average, requirements for fewer working hours. The shaded area shows the average number of overtime hours worked by the crew for the various crew sizes simulated. As expected overtime was reduced as crew members were added and the figure shows a leveling off with a 61-man crew. This indicates comparatively little value in adding additional crew members from the point of view of overtime reduction. The 61-man crew, with a nominal 8-hour workday, worked an average of 7.0 hours plus 0.3 hour overtime; the same size crew on a 12-hour working day worked an average 7.4-hour day, with less than 0.05 hour overtime per day, on the average. Thus comparatively little additional productive work resulted from the 12-hour day since there was little, if any, additional work to do. When an analysis of average regular working hours by classifications and specialty was made, it was shown that crew size increases resulted in less need to work crew members in their alternate specialties, as was expected. About 0.4 hour per day per man was spent in work in a secondary specialty for an 8-hour workday and about 0.2 for a 12-hour day.

COMPLETION OF NONESSENTIAL ACTION UNITS

The number of nonessential action units performed as a function of crew size simulated showed the logically expected trend—more nonessential action units are performed as the crew size increases. A large increase in performance of nonessential action units was also noted as the working hours were increased. For a 12-hour day sufficient work time was available to the crew so that only about 1 percent of the nonessential action units were postponed.

CREW MORALE AND COHESIVENESS

The simulation results for the crew morale variable showed generally that crew morale tended to increase as crew size increased. This is logical since small crews tended to experience heavy overtime with the resultant lowering of morale.

For smaller crews the trend was clearly toward decreasing cohesiveness as the mission progressed. Only in the cases of the 61-man crews did the final mission cohesiveness value exceed the value at the initiation of the mission. Poorer cohesiveness was shown for the crew that worked longer hours but the end-of-mission values did not differ greatly.

FINAL WORDS

MODEL VALIDITY

It has been contended (Stevens, 1951) that measurement is the assignment of numerals to objects or events according to rules. Thus it may be argued that digital simulation, which follows the rules of a program, provides a measure of the degree to which a man-machine system will meet its objectives before the system is actualized. The salient question of the validity of the measurement has been answered for both the group simulation model and previously for the one- and two-operator man-machine simulation model by the data presented in regard to the agreement between the model and outside criteria data. However, support for the validity of a model may rest on other grounds. These grounds include both content validity and construct validity. The content validity of a digital simulation model is the extent to which the content of the model includes all factors necessary for simulating the real world. Thus a model's content validity, in a sense, represents its comprehensiveness. It should be noted that the digital computer merely manipulates the model's content and hence the question of content validity has nothing to do with the employment of the computer itself as a tool. Content validity is concerned only with the inclusiveness of the rules and data put into the computer. For most models a quantitative measure of the sufficiency of the content validity of the model will not be feasible. Obviously it is not possible to include every aspect of the real world into a simulation model. Hence a degree of judgment is required in terms of which aspects to simulate and having selected aspects believed most salient, which parts of the selected aspects to include. While it is not feasible to state the content validity of a model quantitatively, it seems that the model builder should be careful to state why he thinks that his model possesses content validity and which aspects of the situation simulated are not included in the model.

The construct validity of a model that purports psychological representation is the extent to which the predictions of the model can be related to psychological constructs. Another way to show that a model possesses construct validity might be to base the model on the results of a factor analysis. The point is that if a model is to be considered to possess construct validity, a set of facts must exist that would tend to support a contention for the existence of the internal system contained in the model.

SYSTEM MODIFICATION

It may be seen that the use of digital simulation may pay rewards by testing the man-machine interactions in a proposed design. "Testing" here probably means evaluation and in the case of a negative evaluation generation of ideas for a design fix and retest. The areas receiving a negative evaluation may receive attention in a number of ways. Quite obviously a reallocation of tasks to men and machines may be indicated. On the other hand other techniques may produce the same end effects. These might include a change in the physical design of certain subsystems. Controls might be designed for easier activation; displays might be made more readable; the procedures for completing a task may be changed so that sequences are more related, more error free, more quickly performed, or more automated. Sometimes a change in the environmental conditions or a rearrangement of operational equipment will help to overcome a weak spot at a man-machine interactive point. Other techniques for decreasing human error may include the provision of alerting or warning devices and the provision of job aids. If task redesign or some other fix through physical methods is not possible, then operator training may compensate for the deficiency. Related to fixes through "psychological" methods may be consideration of various rewards or incentives for error-free performance or an additional supervisor or operator in order to assure a check during phases of high-error potential. No matter what the improvement may be its effects may be evaluated by resimulation and estimation of whether the later design reduces error sufficiently to allow the system to meet its design objectives.

Of course the results of an overdesigned system, which indicate an underloaded or overmanned situation, may be indicated and similar adjustments in the opposite direction to those discussed above would be appropriate for consideration.

THE FUTURE

It should be pointed out that digital modeling is not considered to be the only weapon available for predicting man-machine effectiveness. Various techniques are available and each is appropriate for use under proper conditions. If used appropriately and sensibly, however, digital simulation can offer answers to questions that must be answered early if effective man-machine systems are to be built with minimum retrofit or redesign. The inherent variability of humans, their viability, and their idiosyncratic behavior, make deterministic prediction of their behavior unrealistic. Computer models, which allow for the variable and often erratic behavior of the human, seem more reasonable for the context and problem area that is being addressed. As has been pointed out elsewhere we can pinpoint a target on the moon but economic forecast errors of 15 percent (that is, errors of $5 to $25 billion in the annual gross national product) are not unexpected. The same may be said of human behavior. While digital psychological simulation models do not represent a panacea, they are better than nothing at all and as we have attempted to show possibly much better than might ordinarily be anticipated. We have attempted to point out by example the method of application and content of two models that have been demonstrated to possess some predictive efficiency. Modeling through digital computer employment and, specifically, the digital simulation of human behavior are considered to be still in an infant stage of development. The development of the field will depend on continued progress in related areas including theoretical psychology, mathematical representation, and computer programing methods.

Shapiro and Rogers (1967) have asserted boldly that, "It can be claimed with some validity that the story of man's progress in science and technology is actually the story of his success in the use of analogy and his progress in simulation." Surely, then, the art of digital simulation, having been placed in the hands of the scientist, engineer, and tactician, represents a relatively new and potentially powerful scheme.

It is hoped that in the years to come this technique will truly prove to be another complementary tool for man's technological progress.

REFERENCES

Adams, O. S., and Chiles, W. I., *Human Performance as a Function of the Work-Rest Ratio During Prolonged Confinement.* ASD Technical Report 61–720, November 1961.

Alluisi, E. A., Hall, T. J., and Chiles, W. D., *Group Performance During Four-hour Periods of Confinement.* Report MRL-TDR-62-70, June 1962.

Altman, I., and Terauds, A., *Major Variables of the Small Group Field.* Contract No. AF49(638)-256, AFOSR-TN 60-1207, 1960.

Apostel, L., Towards the formal study of models in the nonformal sciences, in Kazemier, B., and Vuysje, D. (Eds.), *The Concept and the Role of the Model in Mathematics and Natural and Social Sciences.* Netherlands: Reidel, 1961.

Appley, M. H., and Trumbull, R., *Psychological Stress.* New York: Appleton-Century-Crofts, 1968.

Bekey, G. A., and Gerlough, D. L., Simulation, in R. E. Machol (Ed.), *Systems Engineering Handbook.* New York: McGraw-Hill, 1965.

Blackburn, J. M., *Acquisition of Skill: An Analysis of Learning Curves.* I.H.R.B. Report No, 73, 1936,

Cartwright, D., and Zander, A. (Eds.), *Group Dynamics: Research and Theory.* (2nd ed.) Evanston, Ill.: Peterson, 1962.

Chapanis, A., Men, machines and models, *Amer. Psychol.*, 1961, **16**, 113–131.

Cremeans, J. E., The trend in simulation, *Computers and Automation,* January 1968, 44–48.

Crossman, E. R. F. H., *The Measurement of Perceptual Load in Manual Operations.* Unpublished Ph.D. Thesis, Birmingham Univ., 1956.

Crossman, E. R. F. H., A theory of the acquisition of speed-skill, *Ergonomics,* 1959, **2**, 151–166.

Deitsch, M., Field theory in social psychology, in G. Lindzey (Ed.), *Handbook of Social Psychology.* Cambridge, Mass.: Addison-Wesley, 1954.

De Jong, J. R., The effects of increasing skill in cycle time and its consequences for time standards, *Ergonomics,* 1957, **1**, 51–60.

de Sola Pool, I., Simulating social systems, *International Science and Technology,* March 1964, 62–69.

English, H. B., and English, A. A., *A Comprehensive Dictionary of Psychological and Psychoanalytic Terms.* New York: Longmans, Green, 1958.

George, C. E., Some determinants of small group effectiveness, Research memorandum, Human Resources Research Organization, Ft. Benning, Ga., 1962.

Gilchrist, J. S., Social psychology and group processes, *Annual Review of Psychology.* Palo Alto: Annual Reviews, 1959.

Hare, A. P., *Handbook of Small Group Research.* New York: Free Press-Macmillan, 1962.

Harris, H., Mackie, R. R., and Wilson, C. L., *Performance under Stress; A Review and Critique of Recent Studies*. Los Angeles: Human Factors Research, July 1956.

Haythorne, W. H., and Altman, I., Personality factors in isolated environments, in M. Appley, and R. Trumbull (Eds.), *Psychological Stress*. New York: Appleton-Century-Crofts, 1967.

Hull, C. L., *A Behavior System*. New Haven, Conn.: Yale Univ. Press, 1952.

Kahn, H., *Application of Monte Carlo*. Research memorandum. Santa Monica, Cal.: Rand Corp., April 1954.

Lewin, K., Time perspective and morale, in G. Waston (Ed.), *Civilian Morale*. Boston: Houghton, 1942.

Martin, F. F., *Computer Modeling and Simulation*. New York: Wiley, 1968.

Newcomb, T. M., Individual systems of orientations, in S. Koch (Ed.), *Psychology: A Study of Science*. New York: McGraw-Hill, 1959.

Roseborough, M. E., Experimental studies in small groups, *Psychol. Bull.*, 1953, **50**, 275.

Sayre, M., and Crosson, J., *The Modeling of Mind, Computers and Intelligence*. South Bend: Univ. of Notre Dame Press, 1963.

Sells, S. B., *Dimensions of Group Structure and Group Behavior*. Contract No. AF41(657)-323, Technical Report 61-20, 1961a.

Sells, S. B., *Basic Psychology of Group Behavior*. Contract No. AF41(657)-323, Technical Report 61-19, 1961b.

Shapiro, G., and Rogers, M., *Prospects for Simulation and Simulators of Dynamic Systems*. New York: Spartan, 1967.

Siegel, A. I., Wolf, J. J., and Sorenson, R. T., *Evaluation of a One- or a Two-Operator System Evaluative Model through a Controlled Laboratory Test*. Wayne, Pa.: Applied Psychological Services, 1962.

Stotland, E., Cottrell, N. B., and Laing, G., Group interaction and perceived similarity of members, *J. Abn. and Soc. Psychol.*, 1960, **61**, 335–340.

Thibaut, J. W., and Kelley, H. H., *The Social Psychology of Groups*. New York: Wiley, 1959.

Torrance, E. P., A theory of leadership and interpersonal behavior, in L. Petrullo and B. M. Bass (Eds.), *Leadership and Interpersonal Behavior*. New York: Holt, Rinehart, and Winston, 1961.

APPENDIX A

VARIABLES FOR ONE AND TWO-MAN MODEL

Variable		Symbol	Range of Values	(Maximum Value)
Operator number		j	1, 2	
Partner's operator number		j'	1, 2	
Subtask number		i	xxx	300
Type of subtask	Joint	J	x	
	Equipment	E	x	
	Decision	D	x	
	Cyclic	C	x	
Subtask precedence		d_{ij}	xxx	
Time procedence		I_{ij}	xxxxx.x	
Next subtask (success)		$(i,j)_s$	xxx	
Next subtask (failure)		$(i,j)_f$	xxx	
Average execution time		$\bar{l}_{ij}$	xxxxxx.x	
Average standard deviation		$\bar{\sigma}_{ij}$	xxxxxx.x	
Average probability of subtask success		$\bar{p}_{ij}$	x.xxx	1.0
Time remaining (essential)		T_{ij}^E	xxxxxx.x	
(nonessential)		T_{ij}^N	xxxxxx.x	
Goal aspiration		G_{ij}	x.xxx	1.0
Parameters				
Mission time limit		T_j	xxxxxx.x	
Operator stress threshold		M_j	x.xx	5.0
Operator individuality factor		F_j	x.xxx	
Period for cyclic tasks		P_j	xxxxxx.x	
Number of iterations per run		N	xxx	
Time used		T_{ij}^U	xxxxxx.x	
Stress		s_{ij}	x.xxx	5.0
Stress additive		A_{ij}	x.xxx	
Augmented stress		S_{ij}	x.xxx	
Index of cohesiveness		C_{ij}	x.xxx	
Subtask execution time		t_{ij}	xxxxxx.x	
Pseudo random numbers				
Rectangular distribution		$R_{0,1},\ldots$	.xxxxxxxxxx	
Normal distribution		K_{ij}	x.xxxx	
Probability of subtask success		p_{ij}	x.xxx	1.0
Essentiality indicator		E_{ij}	x	

APPENDIX B
FLOW CHARTS FOR GROUP SIMULATION MODEL

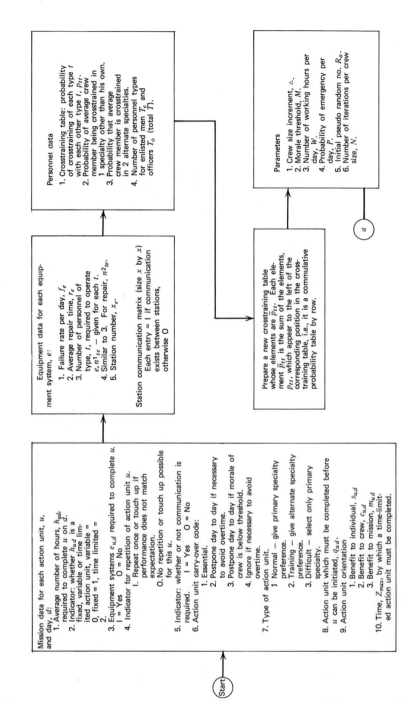

Start

Mission data for each action unit, u, and day, d:
1. Average number of hours, h_{ud}, required to complete u on d.
2. Indicator: whether h_{ud} is a fixed, variable or time limited action unit, variable = 0, fixed = 1, time limited = 2.
3. Equipment systems e_{ud} required to complete u. 1 = Yes 0 = No
4. Indicator for repetition of action unit u.
   1. Repeat once or touch up if performance does not match expectation.
   0. No repetition or touch up possible for this u.
5. Indicator: whether or not communication is required. 1 = Yes 0 = No
6. Action unit carry-over code:
   1. Essential.
   2. Postpone day to day if necessary to avoid overtime.
   3. Postpone day to day if morale of crew is below threshold.
   4. Ignore if necessary to avoid overtime.
7. Type of action unit.
   1 Normal — give primary specialty preference.
   2. Training — give alternate specialty preference.
   3. Difficult — select only primary specialty.
8. Action unit which must be completed before u can be initiated, q_{ud}.
9. Action unit orientation
   1. Benefit to individual, s_{ud}
   2. Benefit to crew, c_{ud}.
   3. Benefit to mission, m_{ud}.
10. Time, Z_{max}, by which a time-limited action unit must be completed.

Equipment data for each equipment system, e:
1. Failure rate per day, f_e.
2. Average repair time, r_e
3. Number of personnel of type, t, required to operate e, $n1_{te}$ — given for each t.
4. Similar to 3. For repair, $n2_{te}$.
5. Station number, x_e.

Station communication matrix (size x by x) Each entry = 1 if communication exists between stations, otherwise 0

Personnel data
1. Crosstraining table: probability of crosstraining of each type t with each other type t, $p_{tt'}$.
2. Probability of average crew member being crosstrained in 1 specialty other than his own.
3. Probability that average crew member is crosstrained in 2 alternate specialties.
4. Number of personnel types for enlisted men T_e and officers T_o (total T).

Prepare a new crosstraining table whose elements are $\bar{p}_{tt'}$. Each element $\bar{p}_{tt'}$ is the sum of the elements, $p_{tt'}$, which appear to the left of the corresponding position in the crosstraining table, i.e., it is a commulative probability table by row.

Parameters
1. Crew size increment, $\triangle$.
2. Morale threshold, M.
3. Number of working hours per day, W.
4. Probability of emergency per day, P.
5. Initial pseudo random no. R_o.
6. Number of iterations per crew size, N.

a

150

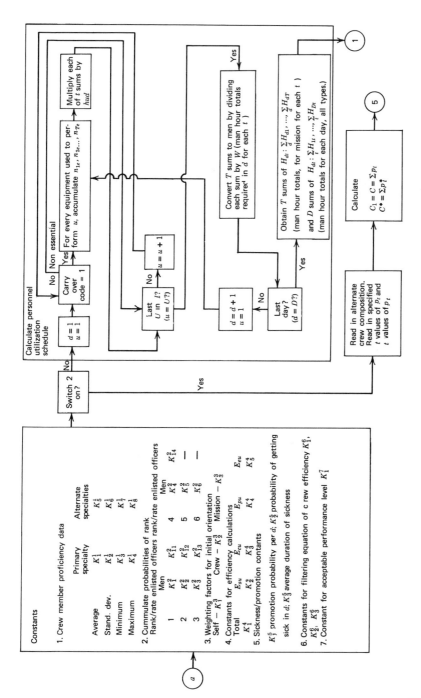

Select initial crew by personnel type

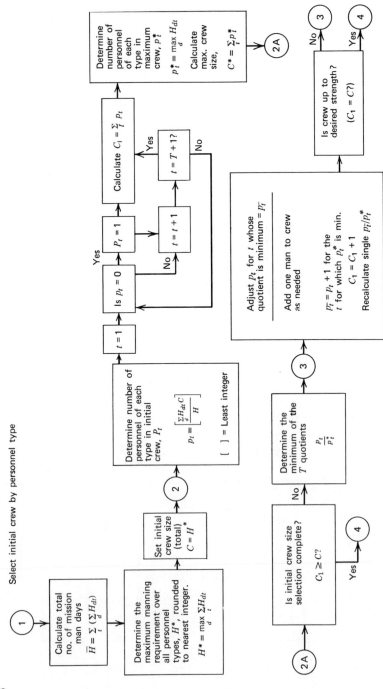

152

Manual adjustment of crew size and composition

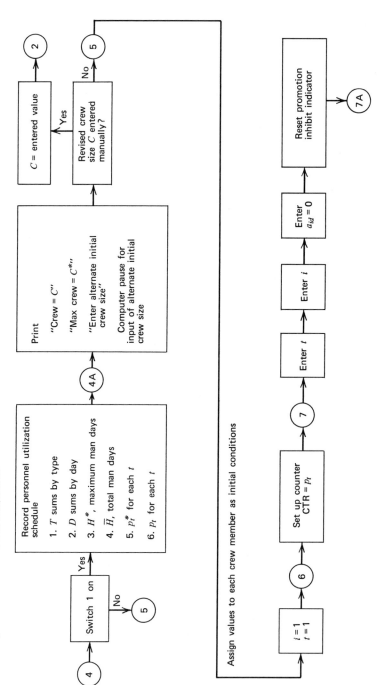

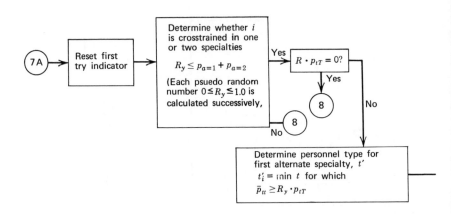

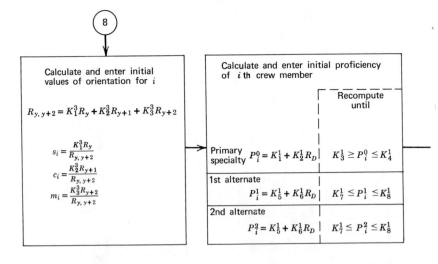

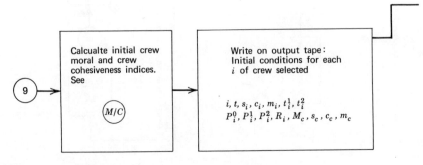

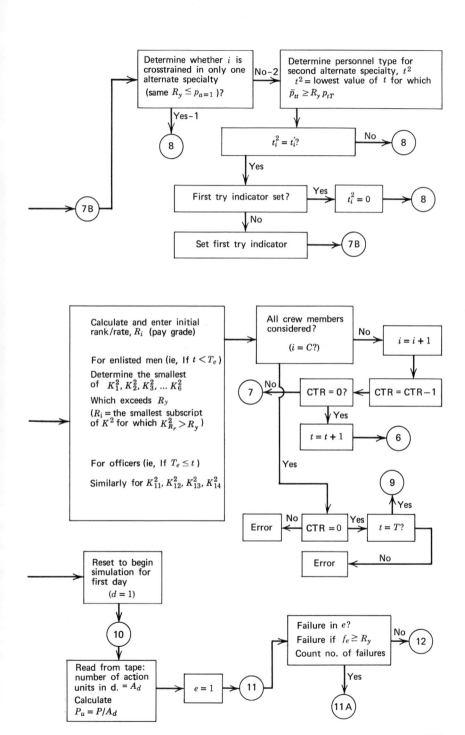

Determine whether i is crosstrained in only one alternate specialty (same $R_y \leq p_{a=1}$)?

No-2 → Determine personnel type for second alternate specialty, t^2
$t^2 =$ lowest value of t for which
$\bar{p}_{tt} \geq R_y\, p_{tT}$

Yes-1 → 8

7B

$t_i^2 = t_i'$? No → 8

Yes

First try indicator set? Yes → $t_i^2 = 0$ → 8

No

Set first try indicator → 7B

Calculate and enter initial rank/rate, R_i (pay grade)

For enlisted men (ie, If $t < T_e$)
Determine the smallest
of $K_1^2, K_2^2, K_3^2, \ldots K_6^2$
Which exceeds R_y
($R_i =$ the smallest subscript
of K^2 for which $K_{R_r}^2 > R_y$)

For officers (ie, If $T_e \leq t$)
Similarly for $K_{11}^2, K_{12}^2, K_{13}^2, K_{14}^2$

All crew members considered?
($i = C$?) No → $i = i + 1$

7 No ← CTR = 0? ← CTR = CTR − 1

Yes

$t = t + 1$ → 6

Yes

9

Yes

Error ← No — CTR = 0 — Yes → $t = T$?

No

Error

Error

Reset to begin simulation for first day
($d = 1$)

10

Read from tape: number of action units in d. $= A_d$
Calculate
$P_u = P/A_d$ → $e = 1$ → 11 →

Failure in e?
Failure if $f_e \geq R_y$
Count no. of failures No → 12

Yes

11 A

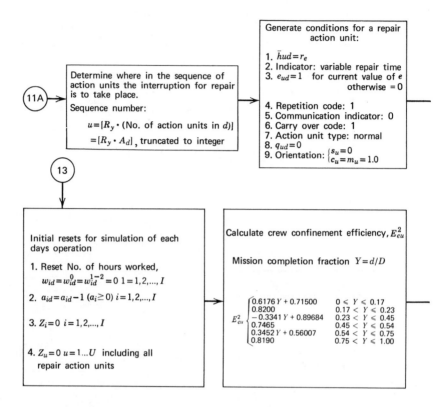

(11A) →

Determine where in the sequence of action units the interruption for repair is to take place.

Sequence number:

$$u = [R_y \cdot (\text{No. of action units in } d)]$$
$$= [R_y \cdot A_d], \text{ truncated to integer}$$

Generate conditions for a repair action unit:

1. $\bar{h}ud = r_e$
2. Indicator: variable repair time
3. $e_{ud} = 1$ for current value of e otherwise $= 0$
4. Repetition code: 1
5. Communication indicator: 0
6. Carry over code: 1
7. Action unit type: normal
8. $q_{ud} = 0$
9. Orientation: $\begin{cases} s_u = 0 \\ c_u = m_u = 1.0 \end{cases}$

(13)

Initial resets for simulation of each days operation

1. Reset No. of hours worked,
$$w_{id} = w_{id}^0 = w_{id}^{1-2} = 0 \quad 1 = 1,2,...,I$$

2. $a_{id} = a_{id} - 1 \ (a_i \geq 0) \ i = 1,2,...,I$

3. $Z_i = 0 \ i = 1,2,...,I$

4. $Z_u = 0 \ u = 1...U$ including all repair action units

Calculate crew confinement efficiency, E_{eu}^2

Mission completion fraction $Y = d/D$

$$E_{eu}^2 \begin{cases} 0.6176\,Y + 0.71500 & 0 \leqslant Y \leqslant 0.17 \\ 0.8200 & 0.17 < Y \leqslant 0.23 \\ -0.3341\,Y + 0.89684 & 0.23 < Y \leqslant 0.45 \\ 0.7465 & 0.45 < Y \leqslant 0.54 \\ 0.3452\,Y + 0.56007 & 0.54 < Y \leqslant 0.75 \\ 0.8190 & 0.75 < Y \leqslant 1.00 \end{cases}$$

Select composition of group which is to accomplish action unit

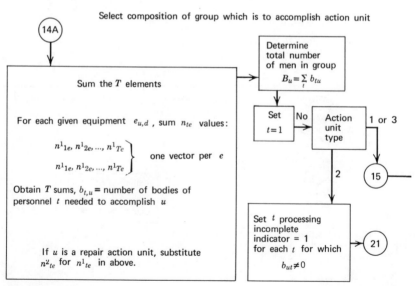

(14A)

Sum the T elements

For each given equipment $e_{u,d}$, sum n_{te} values:

$$\left. \begin{array}{l} n^1{}_{1e}, n^1{}_{2e}, ..., n^1{}_{Tc} \\ n^1{}_{1e}, n^1{}_{2e}, ..., n^1{}_{Te} \end{array} \right\} \text{one vector per } e$$

Obtain T sums, $b_{t,u}$ = number of bodies of personnel t needed to accomplish u

If u is a repair action unit, substitute $n^2{}_{te}$ for $n^1{}_{te}$ in above.

Determine total number of men in group
$$B_u = \sum_t b_{tu}$$

Set $t = 1$ — No — Action unit type — 1 or 3 → **(15)**

| 2

Set t processing incomplete indicator = 1 for each t for which
$$b_{ut} \neq 0$$
→ **(21)**

156

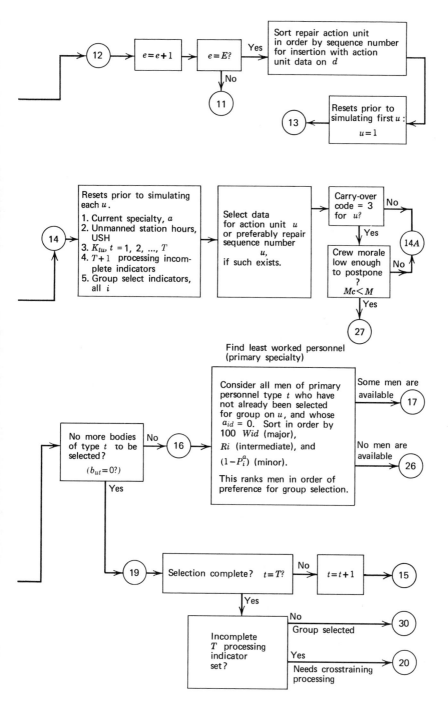

Sort repair action unit in order by sequence number for insertion with action unit data on d

⑫ → $e = e + 1$ → $e = E?$ —Yes→

—No→ ⑪

⑬ ← Resets prior to simulating first u : $u = 1$

⑭ → Resets prior to simulating each u.
1. Current specialty, a
2. Unmanned station hours, USH
3. K_{tu}, $t = 1, 2, ..., T$
4. $T + 1$ processing incomplete indicators
5. Group select indicators, all i

→ Select data for action unit u or preferably repair sequence number u, if such exists.

→ Carry-over code = 3 for u? —No→ ⑭A

—Yes↓

Crew morale low enough to postpone ? $Mc < M$ —No→ ↑

—Yes↓ ㉗

Find least worked personnel (primary specialty)

No more bodies of type t to be selected? $(b_{ut} = 0?)$ —No→ ⑯ →

Consider all men of primary personnel type t who have not already been selected for group on u, and whose $a_{id} = 0$. Sort in order by 100 Wid (major), Ri (intermediate), and $(1 - P_i^a)$ (minor).

This ranks men in order of preference for group selection.

Some men are available → ⑰

No men are available → ㉖

—Yes↓

⑲ → Selection complete? $t = T?$ —No→ $t = t + 1$ → ⑮

—Yes↓

Incomplete T processing indicator set ?

—No / Group selected→ ㉚

—Yes / Needs crosstraining processing→ ⑳

Select group members from primary specialties

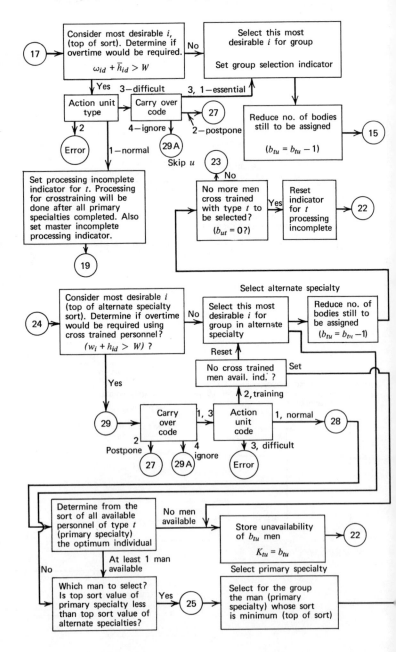

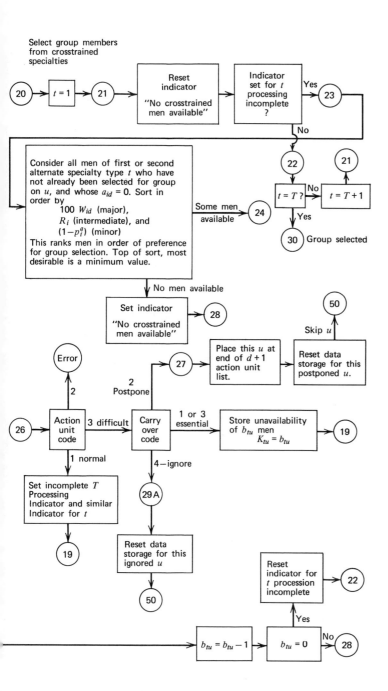

Select group members from crosstrained specialties

20 → t = 1 → 21 → Reset indicator "No crosstrained men available" → Indicator set for t processing incomplete ? → Yes 23

No

22   21

t = T ? → No → t = T + 1

Consider all men of first or second alternate specialty type t who have not already been selected for group on u, and whose a_{id} = 0. Sort in order by
  100 W_{id} (major),
  R_i (intermediate), and
  $(1-p_i^q)$ (minor)
This ranks men in order of preference for group selection. Top of sort, most desirable is a minimum value.

Some men available → 24

Yes

30 Group selected

No men available

Set indicator "No crosstrained men available" → 28

50

Skip u

Error

2

2 Postpone

27 → Place this u at end of d + 1 action unit list. → Reset data storage for this postponed u.

26 → Action unit code → 3 difficult → Carry over code → 1 or 3 essential → Store unavailability of b_{tu} men $K_{tu} = b_{tu}$ → 19

1 normal

4 — ignore

Set incomplete T Processing Indicator and similar Indicator for t

19

29 A

Reset data storage for this ignored u

50

Reset indicator for t procession incomplete → 22

Yes

$b_{tu} = b_{tu} - 1$ → $b_{tu} = 0$ → No 28

159

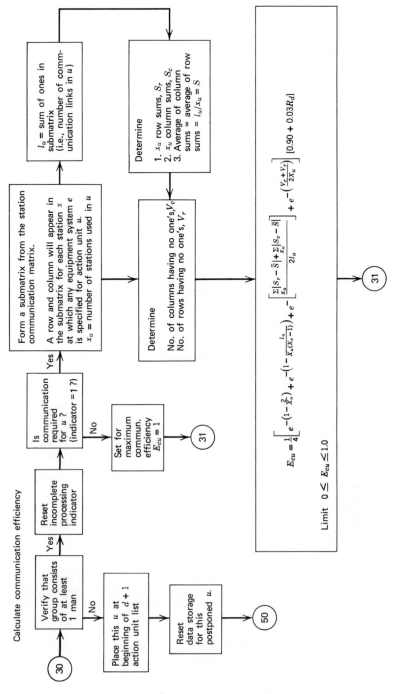

Calculate communication efficiency

160

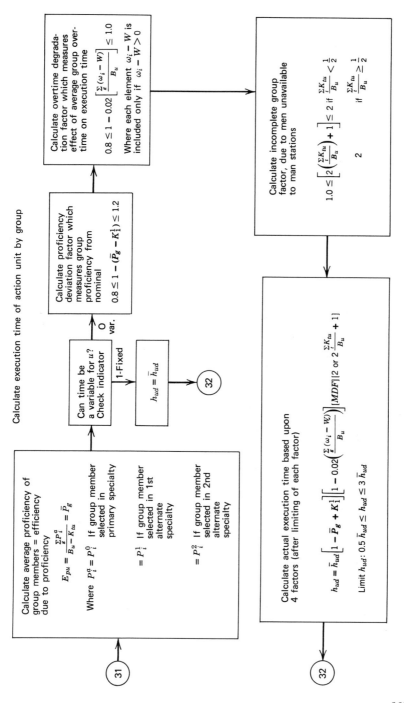

Calculate execution time of action unit by group

Calculate overtime degradation factor which measures effect of average group overtime on execution time

$$0.8 \leq 1 - 0.02 \left[\frac{\sum\limits_{u}(\omega_i - W)}{B_u} \right] \leq 1.0$$

Where each element $\omega_i - W$ is included only if $\omega_i - W > 0$

Calculate proficiency deviation factor which measures group proficiency from nominal

$$0.8 \leq 1 - (\bar{P}_g - K_i^1) \leq 1.2$$

Can time be a variable for u? Check indicator

0 var.

1-Fixed

$h_{ud} = \bar{h}_{ud}$

③②

Calculate average proficiency of group members = efficiency due to proficiency

$$E_{pu} = \frac{\sum\limits_{u} P_i^a}{B_u - K_{tu}} = \bar{P}_g$$

Where $P_i^a = P_i^0$ If group member selected in primary specialty

$= P_i^1$ If group member selected in 1st alternate specialty

$= P_i^2$ If group member selected in 2nd alternate specialty

③①

Calculate incomplete group factor, due to men unavailable to man stations

$$1.0 \leq \left[2\left(\frac{\sum\limits_i K_{tu}}{B_u}\right) + 1\right] \leq 2 \text{ if } \frac{\sum\limits_i K_{tu}}{B_u} < \frac{1}{2}$$

$$2 \quad \text{if } \frac{\sum\limits_i K_{tu}}{B_u} \geq \frac{1}{2}$$

Calculate actual execution time based upon 4 factors (after limiting of each factor)

$$h_{ud} = \bar{h}_{ud}\left[1 - \bar{P}_g + K_i^1\right]\left[1 - 0.02\left(\frac{\sum\limits_u(\omega_i - W)}{B_u}\right)\right]|MDF|\left|2 \text{ or } 2\frac{\sum\limits_i K_{tu}}{B_u} + 1\right|$$

Limit h_{ud}: $0.5\,\bar{h}_{ud} \leq h_{ud} \leq 3\,\bar{h}_{ud}$

③②

161

Adjust records of hours worked for group members

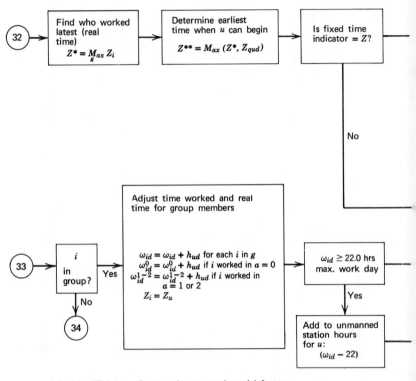

Calculate efficiency of group due to psychosocial factors

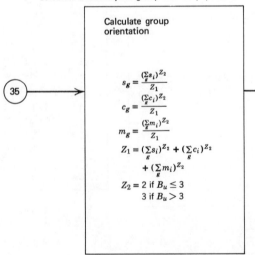

162

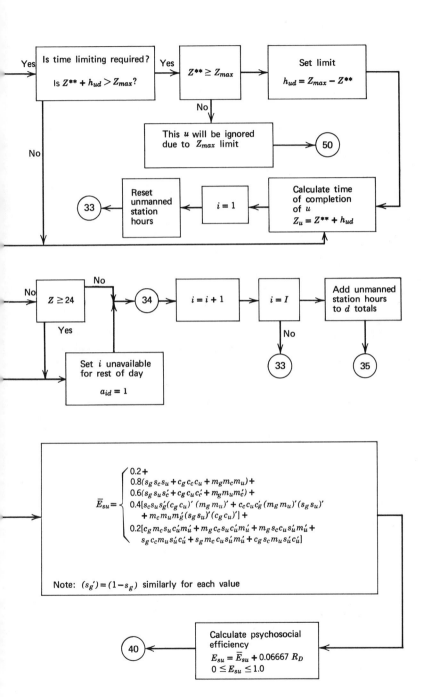

| Is time limiting required?
Is $Z^{**} + h_{ud} > Z_{max}$? | Yes | $Z^{**} \geq Z_{max}$ | | Set limit
$h_{ud} = Z_{max} - Z^{**}$ |

Yes →

No ↓

This u will be ignored due to Z_{max} limit → (50)

Reset unmanned station hours ← $i = 1$ ← Calculate time of completion of u
$Z_u = Z^{**} + h_{ud}$

(33) ←

No

$Z \geq 24$ — No → (34) → $i = i + 1$ → $i = I$ → Add unmanned station hours to d totals

Yes ↓

Set i unavailable for rest of day
$a_{id} = 1$

No ↓ (33)

No ↓ (35)

$$\bar{E}_{su} = \begin{cases} 0.2 + \\ 0.8(s_g s_c s_u + c_g c_c c_u + m_g m_c m_u) + \\ 0.6(s_g s_u s_c' + c_g c_u c_c' + m_g m_u m_c') + \\ 0.4[s_c s_u s_g'(c_g c_u)' \ (m_g m_u)' + c_c c_u c_g'(m_g m_u)'(s_g s_u)' \\ \quad + m_c m_u m_g'(s_g s_u)'(c_g c_u)'] + \\ 0.2[c_g m_c s_u c_u' m_u' + m_g c_c s_u c_u' m_u' + m_g s_c c_u s_u' m_u' + \\ \quad s_g c_c m_u s_u' c_u' + s_g m_c c_u s_u' m_u' + c_g s_c m_u s_u' c_u'] \end{cases}$$

Note: $(s_g') = (1 - s_g)$ similarly for each value

Calculate psychosocial efficiency
$E_{su} = \bar{E}_{su} + 0.06667 R_D$
$0 \leq E_{su} \leq 1.0$ ←

(40) ←

Calculate efficiency

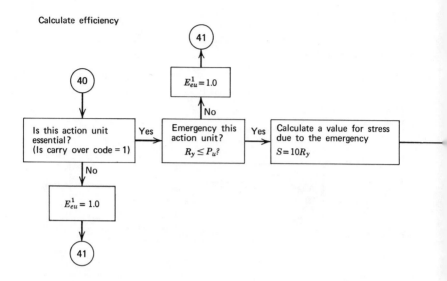

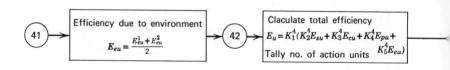

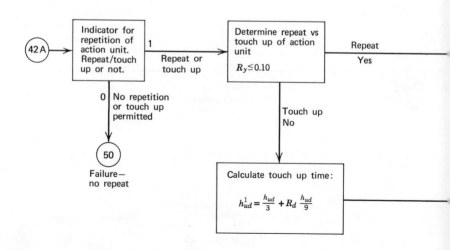

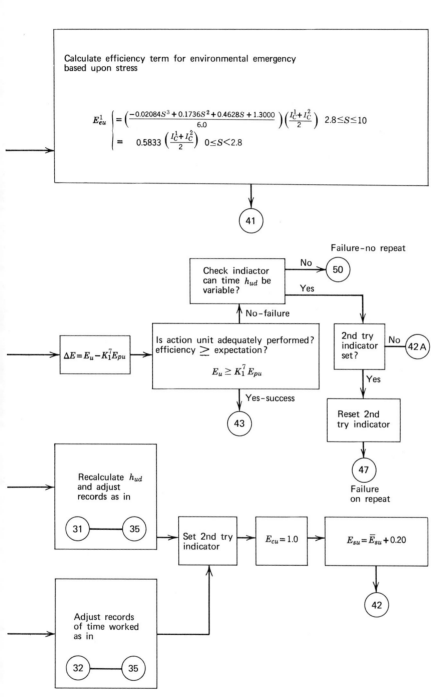

Calculate efficiency term for environmental emergency based upon stress

$$E_{eu}^1 \begin{cases} = \left(\dfrac{-0.02084S^3 + 0.1736S^2 + 0.4628S + 1.3000}{6.0} \right) \left(\dfrac{I_C^1 + I_C^2}{2} \right) & 2.8 \leq S \leq 10 \\[2ex] = \quad 0.5833 \left(\dfrac{I_C^1 + I_C^2}{2} \right) & 0 \leq S < 2.8 \end{cases}$$

41

Failure-no repeat

Check indiactor can time h_{ud} be variable?

No → 50

Yes

No-failure

$\Delta E = E_u - K_1^7 E_{pu}$

Is action unit adequately performed? efficiency $\geq$ expectation?

$$E_u \geq K_1^7 E_{pu}$$

2nd try indicator set?

No → 42 A

Yes

Reset 2nd try indicator

47

Failure on repeat

Yes-success

43

Recalculate h_{ud} and adjust records as in

31 — 35

Set 2nd try indicator

$E_{cu} = 1.0$

$E_{su} = \bar{E}_{su} + 0.20$

42

Adjust records of time worked as in

32 — 35

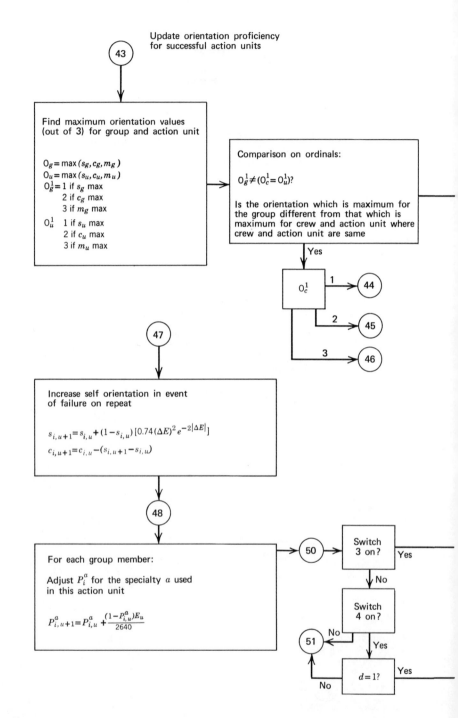

Update orientation proficiency
for successful action units

$\boxed{43}$

Find maximum orientation values
(out of 3) for group and action unit

$O_g = \max(s_g, c_g, m_g)$
$O_u = \max(s_u, c_u, m_u)$
$O_g^1 = 1$ if s_g max
   2 if c_g max
   3 if m_g max

O_u^1 1 if s_u max
   2 if c_u max
   3 if m_u max

Comparison on ordinals:

$O_g^1 \neq (O_c^1 = O_u^1)$?

Is the orientation which is maximum for
the group different from that which is
maximum for crew and action unit where
crew and action unit are same

Yes

O_c^1 1 → $\boxed{44}$

   2 → $\boxed{45}$

   3 → $\boxed{46}$

$\boxed{47}$

Increase self orientation in event
of failure on repeat

$s_{i,u+1} = s_{i,u} + (1 - s_{i,u})[0.74(\Delta E)^2 e^{-2|\Delta E|}]$
$c_{i,u+1} = c_{i,u} - (s_{i,u+1} - s_{i,u})$

$\boxed{48}$

For each group member:

Adjust P_i^a for the specialty a used
in this action unit

$P_{i,u+1}^a = P_{i,u}^a + \dfrac{(1 - P_{i,u}^a)E_u}{2640}$

$\boxed{50}$ → Switch 3 on? Yes

No

Switch 4 on?

$\boxed{51}$ ← No

Yes

$d = 1$? Yes

No

166

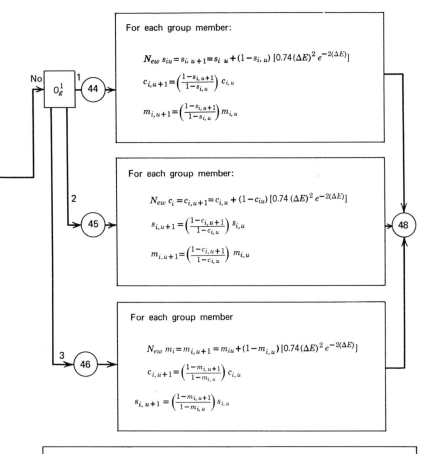

For each group member:

$$New\ s_{iu} = s_{i,u+1} = s_{i\,u} + (1 - s_{i,u})\,[0.74\,(\Delta E)^2\,e^{-2(\Delta E)}]$$

$$c_{i,u+1} = \left(\frac{1 - s_{i,u+1}}{1 - s_{i,u}}\right) c_{i,u}$$

$$m_{i,u+1} = \left(\frac{1 - s_{i,u+1}}{1 - s_{i,u}}\right) m_{i,u}$$

No — 0_g^1 — 1 — 44

For each group member:

$$New\ c_i = c_{i,u+1} = c_{i,u} + (1 - c_{iu})\,[0.74\,(\Delta E)^2\,e^{-2(\Delta E)}]$$

$$s_{i,u+1} = \left(\frac{1 - c_{i,u+1}}{1 - c_{i,u}}\right) s_{i,u}$$

$$m_{i,u+1} = \left(\frac{1 - c_{i,u+1}}{1 - c_{i,u}}\right) m_{i,u}$$

2 — 45

48

For each group member

$$New\ m_i = m_{i,u+1} = m_{iu} + (1 - m_{i,u})\,[0.74\,(\Delta E)^2\,e^{-2(\Delta E)}]$$

$$c_{i,u+1} = \left(\frac{1 - m_{i,u+1}}{1 - m_{i,u}}\right) c_{i,u}$$

$$s_{i,u+1} = \left(\frac{1 - m_{i,u+1}}{1 - m_{i,u}}\right) s_{i,u}$$

3 — 46

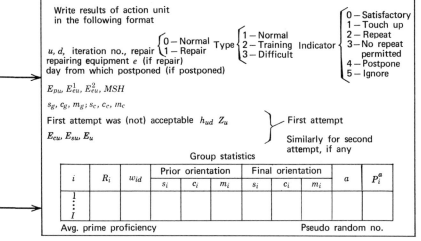

Write results of action unit
in the following format

$u, d,$ iteration no., repair $\begin{cases} 0 - \text{Normal} \\ 1 - \text{Repair} \end{cases}$ Type $\begin{cases} 1 - \text{Normal} \\ 2 - \text{Training} \\ 3 - \text{Difficult} \end{cases}$ Indicator $\begin{cases} 0 - \text{Satisfactory} \\ 1 - \text{Touch up} \\ 2 - \text{Repeat} \\ 3 - \text{No repeat} \\ \quad\ \ \text{permitted} \\ 4 - \text{Postpone} \\ 5 - \text{Ignore} \end{cases}$

repairing equipment e (if repair)
day from which postponed (if postponed)

$E_{pu}, E_{eu}^1, E_{eu}^2, MSH$

$s_g, c_g, m_g; s_c, c_c, m_c$

First attempt was (not) acceptable $h_{ud}\ Z_u$ $\quad$ ⎫ First attempt

E_{cu}, E_{su}, E_u $\quad$ ⎬ Similarly for second
$\quad\quad\quad\quad\quad\quad\quad\quad\quad\quad\quad$ attempt, if any

Group statistics

i	R_i	w_{id}	Prior orientation			Final orientation			a	P_i^a
			s_i	c_i	m_i	s_i	c_i	m_i		
1										
⋮										
I										

Avg. prime proficiency $\qquad\qquad\qquad\qquad$ Pseudo random no.

167

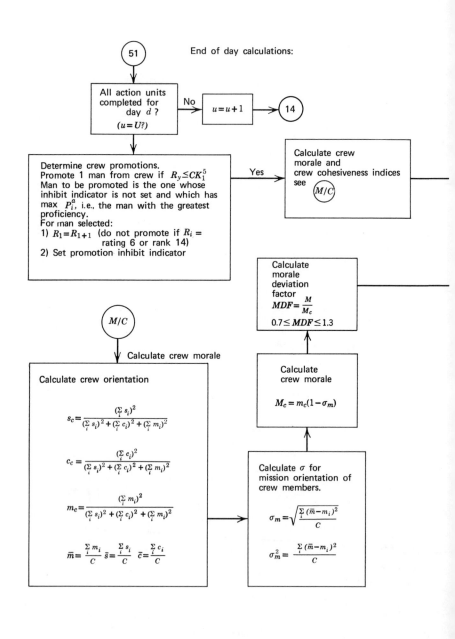

(51)

End of day calculations:

All action units completed for day d? $(u = U?)$

No → $u = u+1$ → (14)

Yes →

Determine crew promotions.
Promote 1 man from crew if $R_y \leq CK_1^5$
Man to be promoted is the one whose inhibit indicator is not set and which has max P_i^a, i.e., the man with the greatest proficiency.
For man selected:
1) $R_1 = R_{1+1}$ (do not promote if $R_i =$ rating 6 or rank 14)
2) Set promotion inhibit indicator

Calculate crew morale and crew cohesiveness indices see (M/C)

Calculate morale deviation factor $MDF = \dfrac{M}{M_c}$
$0.7 \leq MDF \leq 1.3$

(M/C)

↓ Calculate crew morale

Calculate crew orientation

$$s_c = \frac{(\sum_i s_i)^2}{(\sum_i s_i)^2 + (\sum_i c_i)^2 + (\sum_i m_i)^2}$$

$$c_c = \frac{(\sum_i c_i)^2}{(\sum_i s_i)^2 + (\sum_i c_i)^2 + (\sum_i m_i)^2}$$

$$m_c = \frac{(\sum_i m_i)^2}{(\sum_i s_i)^2 + (\sum_i c_i)^2 + (\sum_i m_i)^2}$$

$$\bar{m} = \frac{\sum_i m_i}{C} \quad \bar{s} = \frac{\sum_i s_i}{C} \quad \bar{c} = \frac{\sum_i c_i}{C}$$

Calculate crew morale

$$M_c = m_c(1 - \sigma_m)$$

Calculate σ for mission orientation of crew members.

$$\sigma_m = \sqrt{\frac{\sum_i (\bar{m} - m_i)^2}{C}}$$

$$\sigma_m^2 = \frac{\sum_i (\bar{m} - m_i)^2}{C}$$

(R_y) → $R_{y+1} = 129 R_y + 311715164025_8$ → (Return)

(R_d) → $R_n = \dfrac{1}{8} \sum_{y=1}^{8} R_y$ →

168

Determine sick day additions (all days except last day)

Morale factor $= 1.0 \leq M/Mc \leq 4.0 = MDF$

No. of crew members sick tomorrow $= R_p$ with average $= K_2^5 \cdot C \cdot MDF$

Identity of sick crew members: $i = C \cdot R_y + 1$ (integer part) repeated until all sick men have been identified.

Duration of sickness: $a_{id} = R_p + 1$ with average $= K_3^5$ for each sick man

For each i for which $a_{id} \geq 1$, calculate s_i, c_i, m_i as in (44)

(to increase self orientation) where $\Delta E = (1 - S_i)a_{id}/10$

→ (55)

Calculate crew cohesiveness indices

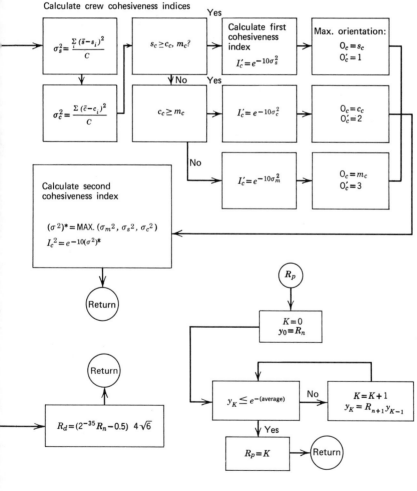

$$\sigma_s^2 = \frac{\sum_i (\bar{s} - s_i)^2}{C}$$

$$\sigma_c^2 = \frac{\sum (\bar{c} - c_i)^2}{C}$$

$s_c \geq c_c, m_c?$

$c_c \geq m_c$

Yes → Calculate first cohesiveness index $I_c' = e^{-10\sigma_s^2}$ → Max. orientation: $O_c = s_c$, $O_c' = 1$

No ↓ Yes → $I_c' = e^{-10\sigma_c^2}$ → $O_c = c_c$, $O_c' = 2$

No → $I_c' = e^{-10\sigma_m^2}$ → $O_c = m_c$, $O_c' = 3$

Calculate second cohesiveness index

$(\sigma^2)^* = MAX. (\sigma_m^2, \sigma_s^2, \sigma_c^2)$

$I_c^2 = e^{-10(\sigma^2)^*}$

(Return)

(Return)

$R_d = (2^{-35} R_n - 0.5)\, 4\sqrt{6}$

(R_p)

$K = 0$
$y_0 = R_n$

$y_K \leq e^{-(average)}$

No → $K = K + 1$
$y_K = R_{n+1} y_{K-1}$

Yes ↓

$R_p = K$ → (Return)

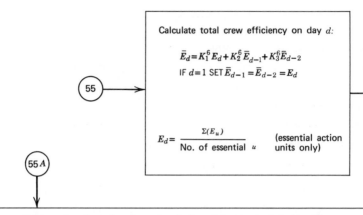

(55)

Calculate total crew efficiency on day d:

$$\bar{E}_d = K_1^6 E_d + K_2^6 \bar{E}_{d-1} + K_3^6 \bar{E}_{d-2}$$

IF $d=1$ SET $\bar{E}_{d-1} = \bar{E}_{d-2} = E_d$

$$E_d = \frac{\Sigma(E_u)}{\text{No. of essential } u} \quad \text{(essential action units only)}$$

(55A)

Write results of one iteration of an entire mission in the following format:
Iteration number, mission duration D, crew size C.
Crew effic. $\bar{E}_d$, crew morale M_c, cohesiveness indices I_c^1, I_c^2
Avg hrs worked $\overline{W}_d$, avg O/T hrs., avg unused hrs, avg hrs/repair
MSH, no. of sick—days, W, M, men unavailable (count of $a \neq 0$)
Avg hrs worked/day (prime, alter., total) crew orientation (s_c, c_c, m_c).

Personnel type	Avg hrs worked/day	No. of men p_t
1 ⋮ T		

Individual statistics

i	R_i	Promotion	Orientation s_i c_i m_i	Proficiencies Prime	Alt -1	Alt -2
I ⋮ T						

Pseudo random no.   Average

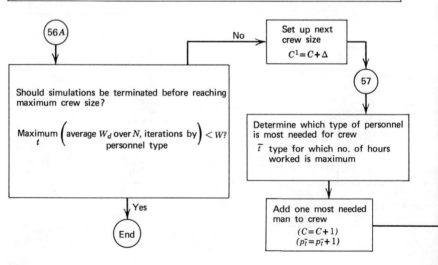

(56A)

Should simulations be terminated before reaching maximum crew size?

$$\underset{t}{\text{Maximum}} \left(\begin{array}{c} \text{average } W_d \text{ over } N, \text{ iterations by} \\ \text{personnel type} \end{array} \right) < W?$$

No → Set up next crew size $C^1 = C + \Delta$

(57)

Determine which type of personnel is most needed for crew
$\bar{t}$ type for which no. of hours worked is maximum

Add one most needed man to crew
$(C = C+1)$
$(p_{\bar{t}} = p_{\bar{t}} + 1)$

Yes ↓

End

Write results of day d in the following format:
d, iteration_number
Crew effic. $\overline{E}_d$, crew morale M_c, cohesiveness indices I_c^1, I_c^2
Avg hrs worked $= \overline{W}_d = w_{id}/C$, avg o/$T$ hrs $W - \overline{W}_d$, avg unused hrs, avg hrs/repair
Total no. of action units, total MSH, crew orientation s_c, c_c, m_c

Repair		Essential		Non-essential		Postponed	Ignored
Satis	Unsatis	Satis	Unsatis	Satis	Unsatis		

Individual statistics

i	R_i	Promotion blank—No 1 —Yes	Orientation			Proficiency			Hrs worked		Type	Days unavail
			s_i	c_i	m_i	Prime	Alt -1	Alt -2	Prime	Alts	t	a_{id}

Mission complete? $d = D$? — Yes → 55 A

No

$d = d+1$ → 10

56 → Completed N iterations for crew size $= C$

No

Increase iteration count by one → 5

Yes

Write results of N iterations of a mission in the following format:
No. of iterations N, D, C
Avg $\overline{E}_d$, avg M_c avg I_c^1 avg I_c^2
Avg $\overline{W}_d$, avg O/T hrs, avg unused hrs, avg hrs/repair
Avg hrs worked/day (prime, alt, total),
Avg s_c, c_c, m_c

Personnel type	Avg hrs worked per day $\overline{\overline{W}}_t$	No. of men p_t
i $\vdots$ I		

Avg crew proficiency P_i^0 Pseudo random no.

(Averages are taken over N iterations.)

Adjust $\overline{\overline{W}}_i$
$$\overline{\overline{W}}_i = \overline{\overline{W}}_i \left(\frac{P_i - 1}{P_i} \right)$$

Crew up to next simulation size? $C = C^1$ — No → 57

Yes

Reset iteration number

Switch on? — No → 5

Yes

Print P_t for each t → 4 A Go to simulate next crew size

56 A ← No — Is crew size maximum $C = C^*$ — Yes → End

171

APPENDIX C

VARIABLES FOR GROUP SIMULATION MODEL

Variable	Symbol	Range of Values (Maximum Value)
Day number	d	xxx
Limit for day number of days	D	(365)
Action unit number	u	xxx
Limit for action units per day	U	(200)
Equipment system number	e	xx
Limit for equipment systems numbers	E	(35)
Personnel type	t	xx
Limit for number of personnel types	$T = T_e + T_0$	(30)
Individual in crew	i	xxx
Crew size	C	(150)
Station number	x	xx
Limit for number of stations	X	(20)
Average performance time (hours)	$\bar{h}_{ud}$	xx.x
Action unit that must be completed before starting u	q_{ud}	xxx
Last real time worked by individual	Z_i	xx.x
Last real time worked on action unit u	Z_u	xx.x
Equipment failure rate per day	f_e	.xxxxx
Repair time	r_e	xx.x
Shift time limit	$Z_{\max}$	xx.x
Carry-over code	—	1, 2, 3, 4
Number of personnel needed to operate	$n_{te}{}^1$	x
Number of personnel needed to repair	$n_{te}{}^2$	x
Number of personnel of type t in crew	p_t	xx
Number of personnel of type t in maximum crew	p_t^*	xx
Number of staff types	T_0	xx
Number of line types	T_e	xx
Predicted man days to be worked	H_{dt}	xxxxxx

Appendix C (*continued*)

Variable	Symbol	Range of Values (Maximum Value)
Predicted total man days	$\bar{H}$	xxxxxx
Predicted maximum man days	H^*	xxxxxx
Probability of cross-training	p_{tt}	.xx
Cumulative probability of cross-training	$\bar{p}_{tt}$	x.xx
Maximum crew size	C^*	xxx
Probability of single cross-training	p_{a-1}	.xxx
Probability of double cross-training	p_{a-2}	.xxx
Actual action unit performance time	h_{ud}	xx.x
Actual touch up time	h'_{ud}	xx.x
Personnel type, alternate specialties	$t_i{}^a$	xx
Proficiency of ith crew member, alternate specialty a	$P_i{}^a$	x.xxx
Probability of an emergency per day	P	.xxxxx
Probability of an emergency per action unit	P_u	.xxxxx
Hours worked by individual, $a = 0$	w_{id}	xx.x
Total hours worked by individual, $a = 1, 2$	$w_{id}{}^{1,2}$	xx.x
Total hours worked, any a	w_{id}	xx.x
Self-orientation of i	s_i	x.xxx
Crew orientation of i	c_i	x.xxx
Mission orientation of i	m_i	x.xxx
Individuals benefit of action unit	s_u	x.xxx
Crew benefit of action unit	c_u	x.xxx
Mission benefit of action unit	m_u	x.xxx
Group	g	
Number of void rows of communications submatrix	V_r	xx
Number of void colmuns of communications submatrix	V_c	xx
Number of stations involved in u	x_u	xx
Number of communication links in u	l_u	xxx
Standard deviation of crew mission orientation	σ_m	.xxxx
Number of men lacking type t	K_{tu}	xx
Rank or rate of i	R_i	(20)
Morale threshold (0 to 1)	M	x.xxx
Crew morale	M_c	x.xxx
Bodies required, type t	b_{tu}	xx
Normal number of working hours per day	W	xx.x
Number of days before i available	a_{id}	xx
Psychosocial efficiency for u	E_{su}	x.xxxx
Average psychosocial efficiency for u	$\bar{E}_{su}$	x.xxxx
Communications efficiency for u	E_{cu}	x.xxxx

Variable	Symbol	Range of Values (Maximum Value)
Efficiency of proficiency (average group proficiency)	E_{pu}	x.xxxx
Number of bodies in group for u	B_u	xx
Efficiency due to environment	E_{eu}	x.xxxx
First cohesiveness index	$I_c{}^1$	xx.xx
Second cohesiveness index	$I_c{}^2$	xx.xx
Group self-orientation	s_g	.xxxx
Group crew orientation	c_g	.xxxx
Group mission orientation	m_g	.xxxx
Crew self-orientation	s_c	.xxxx
Crew orientation	c_c	.xxxx
Crew mission orientation	m_c	.xxxx
Random number uniformly distributed (0 to 1)	R_y	.xxxxxxxxxxx
Initial random number	R_0	.xxxxxxxxxxx
Random deviate, normally distributed	R_d	x.xxx
Poisson distributed random number	R_p	x.xx
Efficiency of g doing u = total efficiency	E_u	.xxxx
Crew efficiency on d	E_d	.xxxx
Smoothed E_d	$\bar{E}_d$	.xxxx
Average hours worked, d	W_d	xx.xx
Crew size increment	Δ	xx
Performance deviation from expectation	ΔE	.xxxx
Manpower shortage hours	MSH	xxxx.x
Morale deviation factor	MDF	x.xxxx
Number of action units in d	A_d	200
Number of iterations per run	N	xxx
Percent mission complete	Y	.xxx
Average hours worked by type t over N simulations	$\overline{\overline{W}}_t$	xx.xx
Alternate specialty	a	x
primary		0
first alternate		1
second alternate		2
Constants for proficiency	$K_1{}^1 - K_8{}^1$	x.xx
Constants for rank	$K_1{}^2 - K_6{}^2, K_{11}{}^2 - K_{14}{}^2$	x.xx
Constants for orientation	$K_1{}^3 - K_3{}^3$	x.xx
Constants for efficiency	$K_1{}^4 - K_5{}^4$	$\pm$x.xxx
Constants for promotion, sickness	$K_1{}^5; K_2{}^5$	.xxxxxx
Constants for sickness duration	$K_3{}^5$	x.xx
Constants for crew efficiency filtering	$K_1{}^6 - K_3{}^6$	$\pm$x.xx
Constant for acceptable performance	$K_1{}^7$	x.xx

INDEX

Date Due

N

I

JL

M.

A

F